NOVEL POROUS MEDIA FORMULATION FOR MULTIPHASE FLOW CONSERVATION EQUATIONS

William T. Sha first proposed the porous media formulation in an article in *Nuclear Engineering and Design* in 1980, and later on with many improvements renamed it the novel porous media formulation (NPMF). The NPMF represented a new, flexible, and unified approach to solving real world engineering problems. Sha introduced a new concept of directional surface porosities and incorporated spatial deviation into the decomposition of all point dependent variables into the formulation. The former greatly improved resolution and modeling accuracy, and the latter made it possible to evaluate all interfacial integrals. A set of conservation equations of mass, momentum, and energy for multiphase flows via time-volume averaging has been rigorously derived for the first time. These equations are in differential-integral form, in contrast to a set of partial differential equations currently used. The integrals arise due to interfacial mass, momentum, and energy transfer.

Dr. William T. Sha is formerly a senior scientist at Argonne National Laboratory and the former director of the Analytic Thermal Hydraulic Research Program and the Multiphase Flow Research Institute. He has published more than 290 papers in the field of thermal hydraulics. He is the recipient of many awards, including the 2005 Technical Achievement Award from the Thermal Hydraulic Division (THD) of the American Nuclear Society (ANS). The highest award given by the THD, "for many outstanding and unique contributions to the field of two phase flow and nuclear reactor design and safety analyses through the development and application of novel computational technique for analyzing thermal hydraulic behavior and phenomena, the development of NPMF of conservation equations used in the COMMIX code, development of boundary fitted coordinates transformation method used in BODYFIT code." He also received the 2006 Glenn T. Seaborg Medal from ANS "for outstanding contributions in understanding multi-dimensional phenomena of natural circulation and fluid stratification in reactor components and systems during normal and off-normal reactor operating conditions" and the 2007 Samuel Untermyer II Medal from ANS "in recognition of pioneering work in the development of significant improvements in NPMF for multiphase flow with far reaching implications and benefits for water cooled reactor components and systems." Most recently he was given the 2008 Reactor Technology Award from ANS "for outstanding leadership and exceptional technical contribution for the U.S. Department of Energy's Industrial Consortium in developing computer codes for intermediate heat exchangers and steam generators of Liquid Metal Fast Breeder Reactors which are based on the NPMF."

Novel Porous Media Formulation for Multiphase Flow Conservation Equations

William T. Sha

Multiphase Flow Research Institute, Director Emeritus

Argonne National Laboratory

Sha & Associates, Inc., President

CAMBRIDGE
UNIVERSITY PRESS

CAMBRIDGE UNIVERSITY PRESS
Cambridge, New York, Melbourne, Madrid, Cape Town,
Singapore, São Paulo, Delhi, Mexico City

Cambridge University Press
32 Avenue of the Americas, New York NY 10013-2473, USA

Published in the United States of America by Cambridge University Press, New York

www.cambridge.org
Information on this title: www.cambridge.org/9781107630178

First published 2011
First paperback edition 2013

A catalogue record for this publication is available from the British Library

Library of Congress Cataloguing in Publication Data

Sha, William T.
Novel porous media formulation for multiphase flow conservation equations /
William T. Sha.
p. cm
Includes bibliographical references and index.
ISBN 978-1-107-01295-0 (hardback)
1. Multiphase flow – Mathematical models. 2. Conservation laws (Mathematics)
I. Title.
TA357.5.M84S52 2011
532′.56–dc22 2011009810

ISBN 978-1-107-01295-0 Hardback
ISBN 978-1-107-63017-8 Paperback

This book is dedicated to

My Parents
Mr. and Mrs. C. F. Sha, and particularly with great affection to my mother, Yunei Gee Sha, whose love and advice have inspired me to obtain the best education, work hard, and contribute to society.

My Wife
Joanne Y. Sha for understanding that I have been working very hard and have not had much time for her. I am deeply grateful she has helped me for so many years.

My Daughters and Son
Ms. Andrea E. Sha Hunt and her husband, Gregory L. Hunt
Dr. Beverly E. Sha and her husband, Dr. Thomas E. Liao, and granddaughter, Grace A. Liao
Professor William C. Sha and his wife, Shawna Suzuki Sha, and grandsons, Samuel Sha and Walter Sha

My Friends
The late Professors B. T. Chao and S. L. Soo for collaborating tirelessly and working with me for more than 25 years. Their contributions are acknowledged.

Contents

Figures and Table

Figures

Table

Foreword

Dr. William T. Sha's longstanding technical achievements and outstanding contributions in the nuclear reactor field are well known both in the United States and abroad. As the director of the Argonne National Laboratory (ANL), I had the privilege of working with Dr. Sha for more than a decade during which he markedly enhanced the reputation of ANL's international reactor programs as the director of the Analytical Thermal Hydraulic Research Program and Multiphase Flow Research Institute. Over many years, his rare combination of analytical rigor and creative insight allowed him to earn international recognition as a leader in the field of thermal hydraulics in both theoretical formulation and reactor design and safety analysis.

His recent work on novel porous media formulation for multiphase flow conservation equations is the subject of this book. "A set of conservation equations of mass, momentum, and energy for two-phase and multiphase flows via time-volume averaging has been rigorously derived for the first time. These equations are in differential-integral form, in contrast to a set of partial differential equations used currently. The integrals arise due to interfacial mass, momentum, and energy transfer." This is an important discovery that will have far reaching implications for both academic and industrial applications. The recent tragic accident at the Fukushima Nuclear Reactor in Japan, which is a boiling-water reactor involving two-phase or multiphase flows, makes the subject of this book even more timely and important.

I have been most impressed by the depth of Dr. Sha's technical knowledge in the area of thermal hydraulics of nuclear reactors. Most importantly, he has always been at the cutting edge of innovation and shares his knowledge with fellow workers, thus advancing the state of the art of thermal hydraulics. His enthusiasm and zest for technical challenges was amazing. It was a real pleasure to work with him.

Dr. Alan Schriesheim, Director Emeritus
Argonne National Laboratory
Member, National Academy of Engineering

Foreword

When I was the manager of reactor physics in the Westinghouse Atomic Power Division [later called the Pressurized Water Reactor (PWR) Division], Dr. William T. Sha worked for me and was instrumental in our development of the first multi-dimensional integral calculation of nuclear-thermal-hydraulic interaction named THUNDER code for the commercial PWRs. The reactivity feedbacks due to thermal-hydraulics, including local subcooled and bulk boiling, control rod insertion, dissolved boron poison in the moderator, and fuel pellet temperature (Doppler effect) were explicitly accounted for. We were then designing Yankee Rowe, Connecticut Yankee, Edison Volta, and Chooz 1. He was clever, indefatigable, and a great asset in our development of the THUNDER codes (WCAP-7006, 1967) and designing these reactors. Plants based on this design are now found in more than half of the world's nuclear power plants. This code represented a quantum jump in design and performance of PWRs when it was successfully completed in 1967.

Once again, Dr. Sha demonstrates innovation and lays the theoretical foundation to develop the novel porous media formulation for multiphase flow conservation equations. The starting point of the novel porous media formulation is Navier-Stokes equations and their interfacial balance equations; the local-volume averaging is performed first via local-volume-averaged theorems, followed by time averaging. A set of conservation equations of mass, momentum, and energy for multiphase systems with internal structures is rigorously derived via time-volume averaging. This set of derived conservation equations has three unique features: (1) the internal structures of the multiphase system are treated as porous media formulation – it greatly facilitates accommodating the complicated shape and size of the internal structures; (2) the concept of directional surface porosities is introduced in the novel porous media formulation and greatly improves modeling accuracy and resolution; and (3) incorporation of spatial deviation for all point dependent variables make it possible to evaluate interfacial mass, momentum, and energy transfer integrals. The novel porous media formulation represents a unified approach for solving real world multiphase flow problems.

Dr. Wm. Howard Arnold
Retired Vice President of Westinghouse Electric Corp.
Member, National Academy of Engineering
Member, U.S. Nuclear Waste Technical Review Board
(Presidential Appointee)

Foreword

Dr. William (Bill) T. Sha is insightful, inventive, and the epitome of professionalism in his technical work.

I have had extensive contacts with Bill Sha, first at Argonne National Laboratory, and later at the U.S. Nuclear Regulatory Commission's Office of Nuclear Regulatory Research. In the latter capacity, I was charged, following the accident at TMI-2, with developing and executing a plan of reactor safety research focused on severe accidents. A major problem facing us was that of knowing whether, when, and how a badly damaged nuclear reactor core could be cooled by natural convection. The obvious problem was that the coolant flow paths were not readily described analytically, even if we knew precisely what they were. We turned to Bill Sha for help with this problem. The response, in a refreshingly short time, was the COMMIX code.

Dr. Sha is the father of the COMMIX code. The code employs the (then) new porous media formulation pioneered by Dr. Sha. The formulation used concept of volume porosity, directional surface porosities, distributed heat source and sink, and distributed resistance and allowed computational analysis of complex geometries critical to power reactor safety analysis. Both conventional porous media and continuum formulations are subsets of this formulation. Dr. Sha's formulation represents a flexible and unified approach to computational fluid dynamics and heat transfer for solving practical engineering problems. Exemplifying Bill's insight, the COMMIX code has proved to be useful for a wide range of engineering design and analysis problems not limited to reactor safety.

The COMMIX code is widely used in the United States and internationally. It has received great attention because of its unique capabilities and features for analyzing inherently multidimensional phenomena such as fluid stratification, natural circulation, coupling effects between reactor core and upper and lower plenums, and so forth. Many foreign countries such as Germany, France, the United Kingdom, Italy, Finland, Japan, China, and South Korea have requested and adopted the COMMIX code and its formulation.

Dr. Sha and his group had the broad range of knowledge, skill, and inventiveness to solve critical engineering problems and acted on behalf of USNRC for an independent verification of the design and performance of passive containment cooling systems of AP-600 or 1000. His novel porous media formulation for multiphase flow conservation equations made significant contributions in the area of reactor design and safety analysis.

Dr. Charles Kelber
Retired Judge
United States Nuclear Regulatory Commission (USNRC)

Nomenclature

A	Area: A_e is the free-flow area of enveloping surface of local averaging volume υ; A_k is the total interfacial area associated with phase k inside υ
c_υ	Specific heat at constant volume
d	Characteristic length of a dispersed phase
D	Diffusivity; also bubble size D^T_{mk} is the eddy diffusivity for mass transfer of phase k $\tilde{D}^T_{mk}$ is the dispersive diffusivity for mass transfer of phase k D_{uk} is the molecular thermal diffusivity of phase k D^T_{uk} is the turbulent diffusivity for internal energy transfer of phase k
E	Total energy per unit mass, $E = u + (1/2)\underline{U}_k \cdot \underline{U}_k$ ${}^{3i}\langle \underline{E}^T_k \rangle$ is the volume-averaged turbulent total energy flux vector of phase k ${}^{3i}\langle \underline{\tilde{E}}_k \rangle$ is the volume-averaged dispersive total energy flux vector of phase k

	${}^{3i}\langle \tilde{\underline{E}}_k^T \rangle$ is the volume-averaged turbulent, dispersive total energy flux vector of phase k
$\underline{f}$	Field force per unit mass
g	Gravitational acceleration
h	Enthalpy per unit mass, $h = u + P/\rho$
	${}^{3i}\langle \underline{h}_k^T \rangle$ is the volume-averaged turbulent enthalpy flux of phase k
	${}^{3i}\langle \tilde{\underline{h}}_k \rangle$ is the volume-averaged dispersive enthalpy flux of phase k
	${}^{3i}\langle \tilde{\underline{h}}_k^T \rangle$ is the volume-averaged turbulent, dispersive enthalpy flux of phase k
H_{kf}	Mean curvature of interface between phases k and f
$\underline{\underline{I}}$	Unitary tensor
J_E	Heat source per unit volume
$\underline{J}_q$	Heat flux vector
ℓ	Characteristic length of local averaging volume υ
L	Characteristic length of physical system; also observation window width
M	Time-averaged interfacial momentum source per unit volume
$\underline{n}$	Unit outward normal vector
P	Pressure
$\dot{Q}_{kf}$	Rate of fluid–fluid interfacial heat transfer into phase k per unit volume of phase k
$\dot{Q}_{wk}$	Rate of fluid–structure interfacial heat transfer into phase k per unit volume of phase k
$\underline{R}$	Resistance per unit volume exerted by stationary, dispersed solid structures

S	Bubble spacing
t	Time
T	Temperature; also averaging time interval
u	Internal energy per unit mass
	${}^{3i}\langle \underline{u}_k^T \rangle$ is the volume-averaged turbulent internal energy flux of phase k
	${}^{3i}\langle \underline{\tilde{u}}_k \rangle$ is the volume-averaged dispersive internal energy flux of phase k
	${}^{3i}\langle \underline{\tilde{u}}_k^T \rangle$ is the volume-averaged turbulent, dispersive internal energy flux of phase k
$\underline{U}$	Velocity vector with its components u, v, and w
v	Local averaging volume
$\underline{W}$	Interface velocity
x, y, z	Cartesian coordinates; also z is elevation
α	Volume fraction
γ_A	Surface porosity
γ_v	Volume porosity
Γ	Interfacial mass source per unit volume due to phase change
Ξ	Interfacial total energy source per unit volume
Π	Interfacial total enthalpy source per unit volume
Θ	Interfacial total internal energy source per unit volume
ε	Eddy diffusivity
κ	Thermal conductivity
κ^T	Turbulent conductivity
λ	Bulk viscosity
μ	Dynamic viscosity
Λ	Length scale of high-frequency fluctuations

ρ	Density of a phase
σ	Surface tension
τ	Characteristic time
$\underline{\underline{\tau}}_k$	Viscous stress tensor ${}^{3i}\langle \underline{\underline{\tau}}_k^T \rangle$ is the volume-averaged Reynolds stress tensor of phase k ${}^{3i}\langle \underline{\underline{\tilde{\tau}}}_k \rangle$ is the volume-averaged dispersive stress tensor of phase k ${}^{3i}\langle \underline{\underline{\tilde{\tau}}}_k^T \rangle$ is the volume-averaged turbulent, dispersive stress tensor of phase k
Φ	Dissipation function ($\Phi_k = \underline{\underline{\tau}}_k : \nabla, \underline{U}_k$)
ϕ_{Ek}	Scalar total energy function
ϕ_{hk}	Scalar enthalpy function
ϕ_{Pk}	Scalar pressure work function
ϕ_{uk}	Scalar internal energy function
$\phi_{\tau k}$	Scalar viscous dissipation function
ψ	Intensive property
$\underline{\psi}_{mk}$	Mass flux function vector
$\underline{\Psi}_{Pk}$	Pressure work function vector
$\underline{\Psi}_{\tau k}$	Viscous stress work function vector

Superscripts

$(\sim)$	Local spatial deviation
$(')$	Fluctuating quantity
$()^T$	Transport properties associated with high-frequency fluctuation
${}^{o}()$	Constant density
T	Fluctuating, or turbulence-related, quantity

Subscripts

c	Capillarity; also characteristic quantity
f	Phase f
HF	High frequency
i	Interface
k	Phase k
kf	Interface of fluid phases k and f, $A_k = A_{kf} = A_{fk} = A_f$
LF	Low frequency
m	Mass; also mixture
q	Heat
s	Interface
u	Internal energy
w	Wall or solid surface
x, y, z	Components in the x, y, and z directions

Symbols

$\underline{(\ )}$	Vector
$\underline{\underline{(\ )}}$	Tensor
$^{2}\langle\rangle$	Area average, local $^{2m}\langle\rangle$ denotes average over free-flow area for the fluid mixture $^{2i}\langle\rangle$ denotes intrinsic average over free-flow area for a phase
$^{3}\langle\rangle$	Volume average $^{3m}\langle\rangle$ denotes volume average over fluid mixture $^{3i}\langle\rangle$ denotes intrinsic volume average of a phase
$^{t}\langle\rangle$	Time average

Operators

∇	Gradient
$\nabla\cdot$	Divergence
$\nabla,$	Dyad
$(\nabla,)_c$	Conjugate of dyad
∇_{kf}	Surface gradient operator along interface between phases k and f
$\frac{d}{dt_k}$	Substantive time derivative $= \frac{\partial}{\partial t_k} + {}^{3i}\langle \underline{U}_k \rangle_{LF} \cdot \nabla$

Acronyms

CPKI	Interfacial capillary pressure transfer integral for phase k
CPFI	Interfacial capillary pressure transfer integral for phase f
EPYTI	Interfacial enthalpy transfer integral
HTI	Interfacial heat transfer integral
IETI	Interfacial internal energy transfer integral
MTI	Interfacial mass transfer integral
MMTI	Interfacial momentum transfer integral
PTI	Interfacial pressure transfer integral
PWI	Interfacial pressure work integral $(\mathrm{PWI})^{(h)}$ is associated with enthalpy production $(\mathrm{PWI})^{(u)}$ is associated with internal energy production
TETI	Interfacial total energy transfer integral
VDI	Interfacial viscous dissipation integral

VSTI	Interfacial viscous stress transfer integral
VWI	Interfacial viscous stress work integral

All equations referred to in the preceding list are written for phase k. For phase f, it is necessary only to change subscript k to f for the entries that appear in the equations.

Preface

This book marks the culmination of a long and exciting career that began in the early 1960s. After I completed my PhD coursework in nuclear engineering at Columbia University, I joined the Atomic Power Division at Westinghouse Electric Corporation (WAPD), where I participated in the final phase design of the world's first pressurized water reactor (PWR) for production of power, Yankee Rowe. After I worked at WAPD for one year, I was selected as the recipient of the Westinghouse Doctorate Scholarship. I returned to Columbia University to complete my doctoral dissertation and received a doctor of engineering science degree in January 1964. After rejoining WAPD, I was assigned to develop the THUNDER computer code for "an integral calculation of steady-state three-dimensional nuclear–thermal–hydraulics interaction," which consisted of a two-group neutron diffusion code and the THINC, a thermal–hydraulics interaction subchannel code. The detailed documentation of the THUNDER code is described in a WAPD-7006 report dated January 1967, written by William T. Sha and Samuel

M. Hendley and labeled Westinghouse Confidential. We published a short paper on the THUNDER code in the *Transactions of the American Nuclear Society* 8(1) (1965) pp. 203–204, but there were no other external publications. The reactivity feedbacks due to thermal hydraulics, including subcooled boiling and bulk boiling, control rod insertion, dissolved boron poison in the moderator, and local fuel pellet temperature (Doppler effect), were explicitly accounted for. Looking back, I realize that this assignment was very advanced at the time, and I learned a lot. After we completed the THUNDER code, I did some calculations using the code for PWRs, which clearly showed significant benefits in reducing hot channel factors.

While I was working on THUNDER, I carefully examined the governing equations used in the code. I noticed that the Boltzmann transport equation was approximated in the form of two-group diffusion equations for the neutronics part of the code and a set of partial differential equations for conservation mass, momentum, and energy for the thermal hydraulics part used in the THINC subchannel code. I developed a simplistic logic to convince myself that the thermal-hydraulics analysis is much more difficult to converge to a solution (or more nonlinear) than the neutronics (reactor physics) calculation for the following reasons. In general, most engineering applications may be divided into three disciplinary areas: (1) solid mechanics for stress and strain analysis, where the interactions of molecules of solid structures are very strong, but the displacement of these molecules is very little; (2) gas dynamics analogous to the neutronics calculations described here, where the interactions between the gas molecules (or neutrons) are minimal, but the displacement of gas molecules (neutrons) is large; and (3) thermal hydraulics for fluid flow with heat transfer analysis, where the interactions between the molecules of fluids are very strong and the displacement of these fluid molecules is very large. Because of the small displacement of molecules in solid structures and very little interaction between gas molecules (or neutrons) in gas dynamics (reactor physics calculations), one can simplify or linearize the corresponding governing equations for both the solid mechanics and the gas dynamics. For the thermal hydraulics analysis, both the interactions and the displacement of fluid molecules are, respectively, very strong and large. Hence, the fluid mechanics with heat transfer is the most difficult subject. Also, I realized that for the neutron transport, we use Boltzmann's equation with integrals, but not for the conservation of mass, momentum, and energy equations for multiphase flow, which are also transport equations for mass, momentum, and energy. Why do they not have integrals? It took me many years to find the answer. This is one of the reasons I have written this book.

In 1967, I joined Argonne National Laboratory. One reason was clear in my mind: I wanted to devote my time and energy to developing a unified formulation for thermal hydraulics analysis, in general, and for nuclear reactor design and safety analysis, in particular. I knew that subchannel analysis was not the answer. It has many inherent deficiencies: (1) for triangular subchannels, the transverse momentum equations are no longer orthogonal to each other, which is theoretically incorrect; (2) from a computational point of view, the computational mesh setup is determined by the subchannel arrangement, thereby losing flexibility for an optimum interfacing with other reactor components inside a reactor vessel or reactor system such as coupling effects among reactor core, upper and lower plenum, and downcomers of PWRs; and (3) cross-flows between the subchannels represent a challenging problem, especially under abnormal reactor operation. After many years of conducting research and carrying out computational fluid dynamics calculations, I proposed the porous media formulation [W.T. Sha, "An Overview on Rod Bundle Thermal Hydraulic Analysis," *Nuclear Engineering and Design* 62 (1980), pp. 1–24].

The porous media formulation uses the concept of volume porosity, directional surface porosities (in my early porous media formulation, surface permeability was used), distributed resistance, and distributed heat source and sink. The volume porosity is defined as the ratio of volume occupied by the fluid(s) to the control volume, and the directional surface porosities are defined as the ratio of the free-flow surface area to the control surface area in three principal directions. The concept of directional surface porosities is derived naturally through the local volume-averaged conservation equations, which are readily calculable quantities. Most practical engineering problems involve many complex shapes and sizes of solid internal structures whose distributed resistance is impossible to quantify accurately. The concept of directional surface porosities eliminates the sole reliance on empirical estimation of the distributed resistance of complex-shaped structures often involved in the analysis. The directional surface porosities thus greatly improve the resolution and modeling accuracy and facilitate mock-ups of numerical simulation models of real engineering systems. Both the continuum and conventional porous media formulations are subsets of the porous media formulation. Moreover, fluid–structure interactions are explicitly accounted for in this formulation. The porous media formulation thus represented a new, flexible, and unified approach to solving real world engineering problems. For *single phase*, this formulation was implemented in the COMMIX computer program (see Chapter 7 and Ref. [37] of this book), which was adopted and used throughout the world in the reactor community.

In recent years, attention has been placed on extension of the formulations from single-phase to multiphase flow. The incorporation of spatial deviation into the decomposition of all point dependent variables made the formulation possible to evaluate interfacial mass, momentum, and energy transfer integrals. A set of conservation equations of mass, momentum, and energy has been rigorously derived via time-volume averaging for the first time. These equations are in differential-integral form, in contrast to a set of partial differential equations used currently. The integrals arise due to interfacial mass, momentum, and energy transfer. The formulation has been greatly improved in recent years and renamed the novel porous media-formulation for multiphase flow conservation equations.

I received a number of awards for my work in thermal hydraulics, in general, and novel porous media formulation, in particular, including the 1986 Pacesetter Award from Argonne National Laboratory "for outstanding creativity and technical leadership in advancing the state-of-the-art of the computational thermal hydraulics through development of a family of COMMIX Computer Programs." I also received the 2005 Technical Achievement Award from the Thermal Hydraulic Division (THD) of the American Nuclear Society (ANS), the highest award offered by the THD, which cited the following: "for many outstanding and unique contributions to the field of two phase flow and nuclear reactor design and safety analyses through the development and application of novel computational technique for analyzing thermal hydraulic behavior and phenomena. The development of novel porous media formulation of governing conservation equations used in the COMMIX code, development of boundary fitted coordinates transformation method used in BODYFIT code." I was awarded the 2006 Glenn T. Seaborg Medal from ANS "for outstanding contributions in understanding multi-dimensional phenomena of natural circulation and fluid stratification in reactor components and systems during normal and off-normal reactor operating conditions" and the 2007 Samuel Untermyer II Medal from ANS "in recognition of pioneering work in the development of significant improvements in novel porous media formulation for multiphase conservation equations resulting in major advancements in the state-of-the-art of multiphase flow with far reaching implications and benefits for water cooled reactor components and systems." Most recently, I earned the 2008 Reactor Technology Award from the ANS "for outstanding leadership and exceptional technical contribution for the U.S. Department of Energy's Industrial Consortium consisting of General Electric Company, Westinghouse Electric Corporation, Atomic International, Foster Wheeler Development Corporation and Babcock & Wilcox in developing computer codes for Intermediate Heat Exchangers and Steam Generators of LMFBR which are based on his Novel Porous Media Formulation."

The intended goal of this book is to make the novel porous media formulation for *multiphase* flow conservation equations equivalent to the Navier-Stokes equations for a single phase. Specifically, the idea is to establish a universally accepted formulation for multiphase flow conservation equations that will greatly reduce both the effort and the expense of developing constitutive relations. Most important, this formulation will enhance and facilitate our fundamental understanding of multiphase flows. It opens a new challenging area and encourages many promising young engineers and scientists to further advance the state of the art in thermal hydraulics, in general, and the novel porous media formulation for multiphase flows, in particular.

I appreciate hearing readers' feedback. If you discover any errors in this book, please let me know.

Dr. William T. Sha
Multiphase Flow Research Institute, Director Emeritus
Argonne National Laboratory
Sha & Associates, Inc., President
2823 Meyers Road
Oak Brook, IL 60523, USA
March 2011

Acknowledgments

I am grateful to Dr. Gary Leaf for stimulating discussions; Dr. Gail Pieper for her excellent editing of my manuscript; and my students Mr. Shuping Wang, Mr. Zoujun Dai, and Ms. Xiaojia Wang for their contributions in typing, proofreading, and making drawings.

1 Introduction

Many years ago, a generic, three-dimensional, time-dependent COMMIX computer program based on the novel porous media formulation for single phase with multicomponent (see Chapter 7) was developed. The computer program was then adopted throughout the world, and the novel porous media formulation has proven both promising and cost effective for many engineering applications. This book now presents the novel porous media formulation for multiphase flow conservation equations.

Multiphase flows consist of interacting phases that are dispersed randomly in space and in time. It is extremely difficult, if not impossible, to track down the interfaces between dispersed phases of multiphase flows. Turbulent, dispersed, multiphase flows can only be described statistically or in terms of averages [1], a fact that was not recognized during the early development of multiphase flow. Averaging procedures are necessary to avoid solving a deterministic multiboundary value problem with the positions of interfaces being a priori unknown. Additional complications arise from the fact that the flow system of interest often contains stationary and complex, solid, heat-generating and heat-absorbing structures. Although, in principle, the intraphase conservation equations for mass, momentum, and energy, as well as their associated initial and boundary conditions, can be written, the problem is far too complicated to permit detailed solutions. In fact, they are seldom needed in engineering applications. A more realistic approach is to express the essential dynamics and thermodynamics of such a system in terms of local volume-averaged quantities. This may be achieved by applying an averaging process, such as time, volume, or statistical averaging. The present work begins with local volume averaging, followed by time averaging. The whole process is called time-local volume averaging, or time-volume averaging.

1.1 Background information about multiphase flow

In earlier papers, the concept of a generalized porous media formulation was conceived [2]* and the local volume-averaged transport equations for multiphase flow were derived in regions containing stationary, heat-generating and heat-absorbing solid structures [3].* Further time averaging of these equations was presented in Refs. [4,5].* A significant step in the development of these averaged equations was the introduction of the concept of volume porosity and

* Directional surface porosity replaces directional surface permeability in Refs. [2–5].

directional surface porosities* associated with stationary and complex solid structures. For a given local volume v, which consists of volume occupied by the fluid mixture v_m and volume occupied by the stationary and complex structures v_w, the volume porosity is defined as the ratio of v_m/v. The directional surface porosities are defined as the ratio of free-flow surface area to the surface area in three principal directions. This concept greatly facilitates the treatment of flow and temperature fields in anisotropic media and enables computational thermal hydraulic analysis of fluids in a region containing complex structures.

Recently, however, it was noted that certain assumptions introduced in Refs. [4,5] regarding the decomposition of the point values of dependent variables, such as density, velocity, pressure, total energy, internal energy, and enthalpy, could be improved [6]. To this end, this book presents a set of time-volume-averaged conservation equations of mass, momentum, and energy for multiphase systems with stationary and complex solid internal structures. Advantage is taken of the use of volume porosity, directional surface porosities, distributed resistance, and distributed heat source and sink. The concept of directional surface porosities was first suggested by Sha [2], greatly improving resolution and modeling accuracy. The governing conservation equations for multiphase flow derived here by time-volume averaging are subject to the length-scale restrictions inherent in the local volume-averaging theorems [7,8] and the time-scale restrictions prescribed in Refs. [4–6].

Time averaging after local volume averaging is chosen over other forms of averaging for the following three reasons:

1. Local volume averaging has been successfully used in many laminar, dispersed multiphase flows, and local volume averaging theorems are well established. Because we are concerned here with turbulent multiphase flows in general and dispersed flows in particular, it is only natural to take the same conventional approach followed by time averaging.
2. Much of our instrumentation records a space average followed by a time average. A Bourdon tube pressure gauge displays an area-averaged pressure averaged over its response time. A hot wire anemometer gives an area-averaged response as a function of time. With a gamma beam, we have a volume-averaged reading over time. We believe that the dependent variables calculated from time-volume-averaged conservation equations are more simply related to the corresponding variables measured by experiment.
3. Within the framework of generalized multiphase mechanics, Soo [9] first suggested that different ranges of particle sizes, densities, and shapes are treated as different dynamic phases. We note that using Eulerian time averaging from the beginning will remove this distinction of dynamic phases. Simple time averaging leads to fractional residence time of a phase rather than volume fraction of a phase. The fractional residence time of a

phase becomes equal to the physical volume fraction only in the case of one-dimensional uniform motion of incompressible phases. Therefore, local volume averaging must precede time averaging.

Local volume averaging has been used extensively for analyzing porous media flows [7,8,10–27]. Volume averaging leads naturally to volume fraction of phases, whereas a priori time averaging yields their fractional residence time. The thermodynamic properties of a fluid, such as density and specific heat, are cumulative with volume fraction but not with fractional residence time, which becomes identical to volume fraction only in the special case of one-dimensional, incompressible flow (see Chapter 6).

The usual approach for single-phase flow is to define for every dependent variable its time average at each point in space. This is feasible but not applicable for turbulent, dispersed multiphase flow in which interface configurations may be moving randomly with time across the point under consideration.

It is awkward to define time-averaged variables in the immediate neighborhood of the phase interface when the phase interface itself is subject to turbulent fluctuations and may be moving back and forth across a particular point during the period for which the time average is defined [1]. Several disadvantages associated with time averaging as the basis of analysis for multiphase flows were pointed out by Reynolds [28] in his review of Ishii's [29] book. Drew [15] and Ishii [29,30] were among the first to recognize the importance of interfacial mass, momentum, and energy balance for two-phase or multiphase flows.

The intuitive answer is that we can describe these multiphase flows only in some average sense, as though each phase occupied all available space of fluids in local volume averaging. We advocate retaining the usual equations of motion appropriate for each phase but using only local volume or time-volume averages of these equations. Closure relations are required because information is lost in the averaging process. However, the approach is general, limited only by our ability to devise these closure relations [1].

1.2 Significance of phase configurations in multiphase flow

The configuration of phases plays a major role in determining the dynamics of multiphase flows and the concomitant heat and mass transport processes when they occur. This is illustrated in Fig. 1.1 for the two extreme cases of highly dispersed flow and ideally stratified flow, which by definition has a plane interface. Figure 1.1 is largely self-explanatory. Given the defining relation for mixture density ρ_m [Eq. (1.2.1)] and for mixture velocity $\underline{U}_m$ [Eq. (1.2.2)], it is easy to show that $\rho_m \underline{U}_m^2$ and $\sum_k \alpha_k \rho_k \underline{U}_k^2$ are not the same. It is also easy to demonstrate that if the Bernoulli relationship for an ideal mixture in highly dispersed flow is written as Eq. (1.2.5), then that for the individual phase must be given by Eq. (1.2.6). Clearly, the form of the Bernoulli equation depends on the configuration of the phases. The Bernoulli equation for other

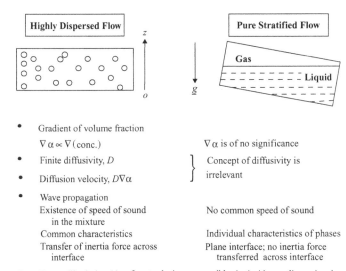

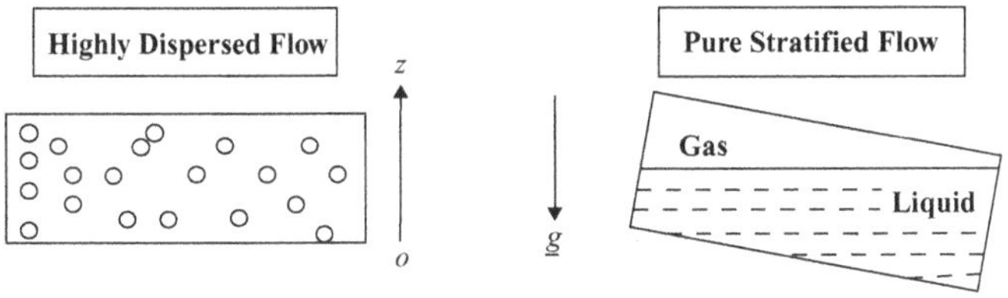

Highly Dispersed Flow	Pure Stratified Flow
• Gradient of volume fraction $\nabla \alpha \propto \nabla(\text{conc.})$	$\nabla \alpha$ is of no significance
• Finite diffusivity, D	Concept of diffusivity is irrelevant
• Diffusion velocity, $D\nabla\alpha$	
• Wave propagation	
Existence of speed of sound in the mixture	No common speed of sound
Common characteristics	Individual characteristics of phases
Transfer of inertia force across interface	Plane interface; no inertia force transferred across interface

- Bernoulli relationships for steady, incompressible, inviscid, one-dimensional flow

$$\rho_m = \sum_k \alpha_k \rho_k, \qquad k = 1,2,\ldots, \tag{1.2.1}$$

where $\alpha_k \rho_k$ is density of phase k based on mixture volume

$$\rho_m U_m = \sum_k \alpha_k \rho_k U_k \tag{1.2.2}$$

Clearly, $$\rho_m U_m^2 \neq \sum_k \alpha_k \rho_k U_k^2 \tag{1.2.3}$$

$$P_m = \sum_k \alpha_k P_k \tag{1.2.4}$$

Ideal Mixture

$$(1/2)\,\rho_m U_m^2 + P_m + \rho_m gz = \text{Constant} \tag{1.2.5}$$

Individual Phase

$$(1/2)\,\rho_k U_k^2 - (1/2)\,\rho_k (U_k - U_m)^2 + P_k + \rho_k gz = \text{Constant} \tag{1.2.6}$$

Fig. 1.1. Significance of phase configurations in multiphase flows.

systems, such as bubbly flow, annular wavy flow with dispersed liquid, intermittent flow, and stratified wavy flow, are far more complex, representing cases intermediate between the highly dispersed flow and ideally stratified flow.

1.3 Need for universally accepted formulation for multiphase flow conservation equations

The current status of multiphase flows with heat transfer is lacking focus and is confusing. The confusion arises from (1) the failure to recognize that the approach to turbulent, dispersed multiphase flows can be described only in terms of averages or statistically; and (2) the fact that because constitutive relations vary with different sets of conservation equations, we have been using many different sets of conservation equations resulting from different formulations such as time averaging, time-volume averaging, volume-time averaging, two-fluid models, and multifluid models. Unless the same set of conservation equations is used, this confusion will persist. The first order of business, then, is to develop a sound, generic, and consistent formulation for rigorously deriving a set of multiphase conservation equations that is equivalent to the Navier-Stokes equations for single phase.

The novel porous media formulation for multiphase flow conservation equations described in this book is the answer. The formulation starts with the local volume averaging of the Navier-Stokes equations and interfacial balance relations of each phase via local volume-averaging theorems that are well established and *theoretically sound*. Time averaging is performed on both the local volume-averaged conservation equations and interfacial balance relations. The time-volume-averaged conservation equations for multiphase flows are in differential-integral form and are in contrast to a set of partial differential equations currently in use. The novel porous media formulation is also *generic*, with the flexibility to include or not the high-frequency fluctuation variables of α'_k, v'_k, and A'_k based on the need from physics of the problem under consideration. Moreover, the formulation is *consistent*: it derives multiphase flow conservation equations based on a single formulation. For example, if we need to retain α'_k and A'_k, then we can follow the same procedure to derive another set of multiphase conservation equations based on the same formulation.

The benefits resulting from the idea of a universally accepted formulation to derive the multiphase flow conservation equations are enormous. The effort and expense in developing constitutive relations no longer need to be repeated for each specific problem. Any developed constitutive relations will deposit in the pool, which can be used for the same or similar problems.

In summary, the unique features of the novel porous media formulation for multiphase flow conservation equations are as follows:

1. The time-volume-averaged multiphase conservation equations are derived in a region that contains stationary and solid internal structures. The fluid–structure interactions are explicitly accounted for via both heat capacity effects during a transient and its additional fluid resistance due to the presence of the structures. This set of time-volume-averaged equations is particularly suitable for numerical analysis with a staggered grid computational system (see Appendix A).

2. Most engineering problems involve many stationary complex shapes and sizes structures whose distributed resistance is impossible to quantify accurately. The concept of directional surface porosities was derived naturally through local volume average; it reduces the sole reliance on an empirical estimation of distributed resistance and provides flexibility to develop a numerical simulation model of a real world engineering system. This concept thus improves resolution and accuracy in modeling results.
3. Introducing spatial deviation of point values of the dependent variables makes it possible to evaluate or approximate interfacial integrals.
4. In deriving the time-volume-averaged continuity equation, the time-volume-averaged interfacial mass generation rate of phase k per unit volume is (see Chapter 5)

$$\gamma_v \alpha_k {}^t\langle \Gamma_k \rangle = -v^{-1} \int_{A_k} {}^t\langle \rho_k (\underline{U}_k - \underline{W}_k) \rangle \cdot \underline{n}_k dA. \quad (1.3.1)$$

This interfacial mass generation rate of phase k per unit volume will directly appear in the time-volume-averaged momentum and energy equations, plus additional interfacial momentum and energy transfer integrals that are intuitively expected (see Chapter 5). It gives some confidence that the derived time-volume-averaged momentum and energy equations for multiphase flows are in good order. We note that Eq. (1.3.1) does not appear in the time averaging formulation.

5. For any single-phase fluid system ($A_k = 0, \alpha_k = 1$) with a constant fluid density without internal structure ($\gamma_A = \gamma_v = 1$), the derived time-volume-averaged continuity equation reduces to $\nabla \cdot \underline{U}_k = 0$, which is as expected (see Chapter 5).
6. The derived time-volume momentum equation reduces to the basic relation of fluid hydrostatics pressure for any single-phase (fluid) system ($\underline{A}_k = 0$, $\alpha_k = 1$) without internal structure ($\gamma_v = \gamma_A = 1$). Therefore, all interfacial integrals vanish. If the fluid is at rest, then all quantities associated with $\underline{U}_k$ also vanish. With $\underline{f} = \underline{g}$, $\underline{g}$ being the gravitational acceleration vector in vertical direction, Eqs. (5.5.7f), (5.5.7g), and (5.5.7j) (see Chapter 5) become

$$-\nabla P_k + \rho_k \underline{g} = 0. \tag{1.3.2}$$

Equation (1.3.2) satisfies the basic hydrostatic pressure of a fluid as expected.

7. Both the continuum and conventional porous media formulations are subsets of the novel porous media formulation [2–6].

Items 4, 5, and 6 give confidence in the derived time-volume-averaged conservation equations (see Chapter 5).

2 Averaging relations

A general flow system occupying a region is illustrated in Fig. 2.1. The flow system coincides with the constant local averaging volume υ, which is invariant in both space and time, and its orientation relative to inertial frame of reference is fixed. It has an enveloping surface of area A with unit outward normal $\underline{n}$. The region consists of a partially and/or totally immersed stationary solid phase w, and a fluid mixture with phase k and other phase or phases f flowing through the region. Phase k has a variable volume υ_k with total interfacial area A_k in υ. A portion of A_k is made of fluid–fluid interface A_{kf}, and the remainder is fluid–solid interface A_{wk}. The unit normal vector $\underline{n}_k$ of A_k is always drawn outward from phase k, regardless of whether it is associated with A_{kf} or A_{wk}. The local velocity of phase k is $\underline{U}_k$ and that of the interface $\underline{W}_k$. On A_{wk}, $\underline{W}_k$ vanishes except when there is a chemical reaction or if the solid is porous and fluid is passing through the pores.

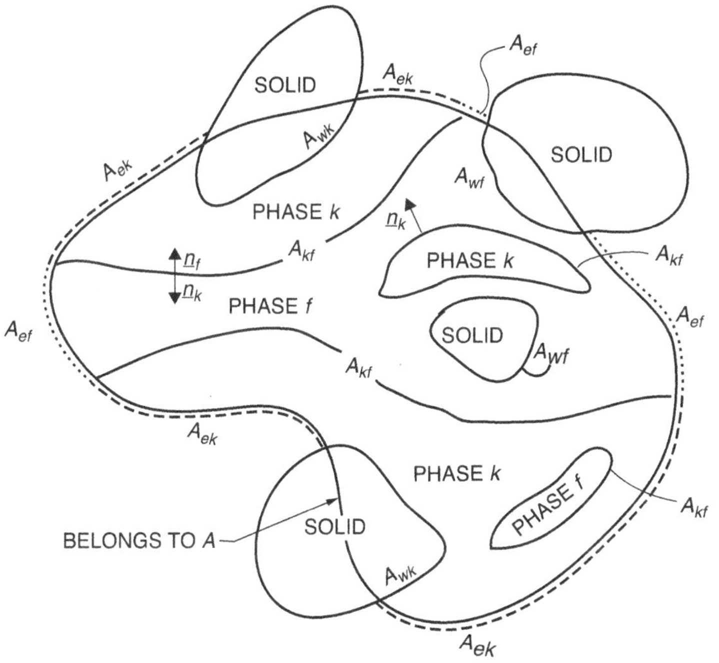

Fig. 2.1. Multiphase flow system with stationary and solid structures (local averaging volume υ with enveloping surface A).

2.1 Preliminaries

Based on Fig. 2.1, the following relations hold:

a. Volume of fluid mixture:

$$\upsilon_m = \sum_k \upsilon_k \tag{2.1.1}$$

b. Local averaging volume:

$$\upsilon = \upsilon_m + \upsilon_w, \tag{2.1.2}$$

where v_w is the total volume of the stationary and complex solid structures in v.

c. Volume porosity:

$$\gamma_v = v_m/v = 1 - (v_w/v), \tag{2.1.3}$$

which is a constant for a given v.

d. Volume fraction of phase k in fluid mixture:

$$\alpha_k = v_k/v_m, \tag{2.1.4}$$

which is a dependent variable.

2.2 Local volume average and intrinsic volume average

For any intensive property ψ_k associated with phase k, be it a scalar, vector, or tensor, the local volume average of ψ_k is defined by

$$^{3}\langle\psi_k\rangle = \frac{1}{v}\int_{v_k}\psi_k dv = \gamma_v\frac{1}{v_m}\int_{v_k}\psi_k dv = \gamma_v\alpha_k\frac{1}{v_k}\int_{v_k}\psi_k dv. \tag{2.2.1}$$

The intrinsic volume and fluid mixture volume averages for phase k are defined, respectively, as follows:

$$^{3i}\langle\psi_k\rangle = \frac{1}{v_k}\int_{v_k}\psi_k dv \tag{2.2.1a}$$

$$^{3m}\langle\psi_k\rangle = \frac{1}{v_m}\int_{v_k}\psi_k dv. \tag{2.2.1b}$$

These averages are related according to

$$^{3}\langle\psi_k\rangle = \gamma_v\,^{3m}\langle\psi_k\rangle = \gamma_v\alpha_k\,^{3i}\langle\psi_k\rangle. \tag{2.2.2}$$

If we set $\psi_k = 1$, then ${}^{3i}\langle 1\rangle = 1$, ${}^{3m}\langle 1\rangle = \alpha_k$, and ${}^{3}\langle 1\rangle = \gamma_v \alpha_k$.

2.3 Local area average and intrinsic area average

The various area averages of ψ_k are defined in a manner similar to that for the volume averages. The local area average of ψ_k is defined by

$$ {}^{2}\langle \psi_k \rangle = \frac{1}{A}\int_{A_{ek}} \psi_k \, dA = \gamma_A \frac{1}{A_e}\int_{A_{ek}} \psi_k \, dA, \quad (2.3.1) $$

where A_e is the total free flow area available for the fluid mixture to enter or exit from the averaging volume v with enveloping surface A, and A_{ek} is that allotted to phase k. The surface porosity γ_A is defined by

$$ \gamma_A = \frac{A_e}{A}, \quad (2.3.2) $$

that is, the fraction of the enveloping surface A through which the fluid mixture flows.

The area average of ψ_k over the total free flow area A_e is .

$$ {}^{2m}\langle \psi_k \rangle = \frac{1}{A_e}\int_{A_{ek}} \psi_k \, dA = \frac{A_{ek}}{A_e}\frac{1}{A_{ek}}\int_{A_{ek}} \psi_k \, dA, \quad (2.3.3) $$

and its intrinsic area average is

$$ {}^{2i}\langle \psi_k \rangle = \frac{1}{A_{ek}}\int_{A_{ek}} \psi_k \, dA. \quad (2.3.4) $$

Clearly,

$$ {}^{2}\langle \psi_k \rangle = \gamma_A {}^{2m}\langle \psi_k \rangle = \gamma_A \frac{A_{ek}}{A_e} {}^{2i}\langle \psi_k \rangle, \quad (2.3.5) $$

where A_{ek}/A_e is the fraction of the free flow area allotted to phase k.

The meaning of A_{ek}/A_e in Eq. (2.3.5) for a homogeneous, nonstructural fluid medium can be seen by examining the mass flux at a bounding surface of a local averaging volume in the form of a rectangular parallelepiped $\Delta x \Delta y \Delta z$. Consider, for example, the mass flow rate of a mixture of two phases k and f through area $\Delta A_{e,x}$, which may be a portion of ΔA_x $(=\Delta y \Delta z)$. Thus, $\Delta A_{e,x} = \gamma_{AX} \Delta A_x$. Clearly,

$$\rho_m U_{mx} \Delta A_{e,x} = \rho_k U_{kx} \Delta A_{ek,x} + \rho_f U_{fx} \Delta A_{ef,x}, \qquad (2.3.6)$$

where U_{mx}, U_{kx}, and so forth, denote velocity components along the x axis. Now for a homogeneous fluid medium, we have

$$\rho_m U_{mx} = \alpha_k \rho_k U_{kx} + \alpha_f \rho_f U_{fx}. \qquad (2.3.7)$$

It follows that

$$\frac{\Delta A_{ek,x}}{\Delta A_{e,x}} = \alpha_k \quad \text{and} \quad \frac{\Delta A_{ef,x}}{\Delta A_{e,x}} = \alpha_f. \qquad (2.3.8a)$$

Using the same reasoning, one may show that for flow in the y and z direction and through $\Delta A_{e,y}$ and $\Delta A_{e,z}$, respectively,

$$\frac{\Delta A_{ek,y}}{\Delta A_{e,y}} = \alpha_k, \quad \text{and} \quad \frac{\Delta A_{ef,y}}{\Delta A_{e,y}} = \alpha_f \qquad (2.3.8b)$$

$$\frac{\Delta A_{ek,z}}{\Delta A_{e,z}} = \alpha_k, \quad \text{and} \quad \frac{\Delta A_{ef,z}}{\Delta A_{e,z}} = \alpha_f. \qquad (2.3.8c)$$

We emphasize that the preceding results are valid for approximating a homogeneous nonstructural fluid medium, as pointed out by Whitaker [7]. Strictly speaking, they are applicable only to highly dispersed systems. The length-scale restrictions of the local volume averaging theorems developed by Whitaker [7] are consistent with these approximations.

2.4 Local volume averaging theorems and their length-scale restrictions

The local volume averages of the spatial and time derivatives of a fluid property ψ_k, which may be a scalar, vector, or tensor, are given by Whitaker [7,10], Slattery [8], Anderson and Jackson [11], Gray and Lee [23], and others. They are related to the corresponding derivatives of the averages and an interfacial area integral according to the following relations:

$$^3\langle\nabla\psi_k\rangle = \nabla\,^3\langle\psi_k\rangle + \upsilon^{-1}\int_{A_k}\psi_k\underline{n}_k\,dA, \qquad (2.4.1a)$$

$$^3\langle\nabla\cdot\underline{\psi}_k\rangle = \nabla\cdot\,^3\langle\underline{\psi}_k\rangle + \upsilon^{-1}\int_{A_k}\underline{\psi}_k\cdot\underline{n}_k\,dA, \qquad (2.4.1b)$$

and

$$^3\left\langle\frac{\partial\psi_k}{\partial t}\right\rangle = \frac{\partial\,^3\langle\psi_k\rangle}{\partial t} - \upsilon^{-1}\int_{A_k}\psi_k\underline{W}_k\cdot\underline{n}_k\,dA. \qquad (2.4.2)$$

In the previous equations, A_k denotes the sum of all interfacial areas associated with phase k inside the local averaging volume υ. Thus, referring to Fig. 2.1, we see that A_k consists of the fluid–fluid interface A_{kf} and the fluid–solid interface A_{wk}. For a stationary, nonporous, and nonreacting solid, $\underline{W}_k$ vanishes on A_{wk}.

We note that these averaging relations are subject to the following length-scale restrictions, first given by Whitaker [7]:

$$d \ll \ell \ll L, \tag{2.4.3}$$

where d is a characteristic length of the dispersed phase, ℓ is a characteristic length of υ, and L is that of the physical system. Therefore, the averaging volume cannot be made arbitrarily small.

Whitaker [7], Slattery [8], and Gray and Lee [23] also showed that

$$\nabla \cdot {}^3\langle \underline{\psi}_k \rangle = \upsilon^{-1} \int_{A_{ek}} \underline{\psi}_k \cdot \underline{n}_k \, dA. \tag{2.4.4}$$

In the Cartesian coordinate system, $\underline{\psi}_k = \underline{i}\psi_{kx} + \underline{j}\psi_{ky} + \underline{k}\psi_{kz}$, where $\underline{i}$, $\underline{j}$, and $\underline{k}$ are unit vectors in the positive x, y, and z directions, respectively. For $\upsilon = \Delta x \Delta y \Delta z$ centered at point (x, y, z), Eq. (2.4.4) can be written as follows:

$$\begin{aligned}
&\nabla \cdot {}^3\langle \underline{\psi}_k \rangle \\
&\cong \frac{1}{\Delta x}\left(\frac{\Delta A_{e,x+(\Delta x/2)}}{\Delta y \Delta z} \frac{\Delta A_{ek,x+(\Delta x/2)}}{\Delta A_{e,x+(\Delta x/2)}} \frac{1}{\Delta A_{ek,x+(\Delta x/2)}} \right. \\
&\quad \times \int_{A_{ek,x+(\Delta x/2)}} \psi_{kx} \, dA_x - \frac{\Delta A_{e,x-(\Delta x/2)}}{\Delta y \Delta z} \frac{\Delta A_{ek,x-(\Delta x/2)}}{\Delta A_{e,x-(\Delta x/2)}} \\
&\quad \left. \times \frac{1}{\Delta A_{ek,x-(\Delta x/2)}} \int_{A_{ek,x-(\Delta x/2)}} \psi_{kx} \, dA_x \right) \\
&\quad + \frac{1}{\Delta y}\left(\frac{\Delta A_{e,y+(\Delta y/2)}}{\Delta z \Delta x} \frac{\Delta A_{ek,y+(\Delta y/2)}}{\Delta A_{e,y+(\Delta y/2)}} \frac{1}{\Delta A_{ek,y+(\Delta y/2)}} \right. \\
&\quad \times \int_{A_{ek,y+(\Delta y/2)}} \psi_{ky} \, dA_y - \frac{\Delta A_{e,y-(\Delta y/2)}}{\Delta z \Delta x} \frac{\Delta A_{ek,y-(\Delta y/2)}}{\Delta A_{e,y-(\Delta y/2)}}
\end{aligned}$$

$$\times \frac{1}{\Delta A_{ek,y-(\Delta y/2)}} \int_{A_{ek,y-(\Delta y/2)}} \psi_{ky} \, dA_y \Bigg)$$

$$+ \frac{1}{\Delta z} \Bigg(\frac{\Delta A_{e,z+(\Delta z/2)}}{\Delta x \Delta y} \frac{\Delta A_{ek,z+(\Delta z/2)}}{\Delta A_{e,z+(\Delta z/2)}} \frac{1}{\Delta A_{ek,z+(\Delta z/2)}}$$

$$\times \int_{A_{ek,z+(\Delta z/2)}} \psi_{kz} \, dA_z - \frac{\Delta A_{e,z-(\Delta z/2)}}{\Delta x \Delta y} \frac{\Delta A_{ek,z-(\Delta z/2)}}{\Delta A_{e,z-(\Delta z/2)}}$$

$$\times \frac{1}{\Delta A_{ek,z-(\Delta z/2)}} \int_{A_{ek,z-(\Delta z/2)}} \psi_{kz} \, dA_z \Bigg)$$

$$\cong \frac{\partial}{\partial x} \gamma_{Ax} \alpha_k{}^{2i} \langle \psi_{kx} \rangle + \frac{\partial}{\partial y} \gamma_{Ay} \alpha_k{}^{2i} \langle \psi_{ky} \rangle + \frac{\partial}{\partial z} \gamma_{Az} \alpha_k{}^{2i} \langle \psi_{kz} \rangle. \tag{2.4.5}$$

Here, γ_{Ax}, γ_{Ay}, and γ_{Az} are directional surface porosities defined by

$$\gamma_{Ax} = \begin{cases} \gamma_{A,x+(\Delta x/2)} = \dfrac{\Delta A_{e,x+(\Delta x/2)}}{\Delta y \Delta z} \\ or \\ \gamma_{A,x-(\Delta x/2)} = \dfrac{\Delta A_{e,x-(\Delta x/2)}}{\Delta y \Delta z} \end{cases} \tag{2.4.6a}$$

$$\gamma_{Ay} = \begin{cases} \gamma_{A,y+(\Delta y/2)} = \dfrac{\Delta A_{e,y+(\Delta y/2)}}{\Delta z \Delta x} \\ or \\ \gamma_{A,y-(\Delta y/2)} = \dfrac{\Delta A_{e,y-(\Delta y/2)}}{\Delta z \Delta x} \end{cases} \tag{2.4.6b}$$

$$\gamma_{Az} = \begin{cases} \gamma_{A,z+(\Delta z/2)} = \dfrac{\Delta A_{e,z+(\Delta z/2)}}{\Delta x \Delta y} \\ or \\ \gamma_{A,z-(\Delta z/2)} = \dfrac{\Delta A_{e,z-(\Delta z/2)}}{\Delta x \Delta y}. \end{cases} \tag{2.4.6c}$$

For compactness, we write Eq. (2.4.5) in vectorial form as

$$\nabla \cdot {}^{3}\langle \underline{\psi}_{k} \rangle = \nabla \cdot \gamma_{A} \alpha_{k} {}^{2i}\langle \underline{\psi}_{k} \rangle. \tag{2.4.7}$$

Equation (2.4.7) is used for all flux-related quantities in the governing time-volume-averaged conservation equations. It is strictly valid only for homogeneous dispersed systems.

Upon setting $\psi_k = 1$ in Eq. (2.2.2), one obtains ${}^{3}\langle 1 \rangle = \gamma_{v}\alpha_{k}$; hence, Eq. (2.4.2) gives

$$\gamma_{v} \frac{\partial \alpha_{k}}{\partial t} = v^{-1} \int_{A_k} \underline{W}_{k} \cdot \underline{n}_{k} \, dA \tag{2.4.8}$$

because γ_v is time independent. Furthermore, Eq. (2.4.1a) gives

$$\nabla \gamma_{v} \alpha_{k} = -v^{-1} \int_{A_k} \underline{n}_{k} \, dA. \tag{2.4.9}$$

The physical interpretation of Eq. (2.4.9) is shown in Appendix B with $\gamma_v = 1$.

To conclude this section, we reiterate that, on the one hand, for flux-related quantities, $\underline{\psi}_k$, we have the vectorial form demonstrated in Eq. (2.4.7). On the other hand, for a vector $\underline{\psi}_k$ that is not flux related, we have

$$\nabla \cdot {}^{3}\langle \underline{\psi}_{k} \rangle = \nabla \cdot \gamma_{v} \alpha_{k} {}^{3i}\langle \underline{\psi}_{k} \rangle. \tag{2.4.10}$$

For any scalar-intensive properties ψ_k, the corresponding relation is

$$\nabla {}^{3}\langle \psi_{k} \rangle = \nabla \gamma_{v} \alpha_{k} {}^{3i}\langle \psi_{k} \rangle. \tag{2.4.11}$$

2.5 Conservative criterion of minimum size of characteristic length of local averaging volume

To apply the local volume-averaging theorems outlined in Section 2.4, the characteristic length-scale restriction shown in Eq. (2.4.3) must be observed. We require the characteristic length of local averaging volume (l) to be sufficiently large that averages over any dispersed phase inside υ enveloped by surface A, as shown in Fig. 2.1, vary smoothly with position. Whenever possible, the characteristic length of l should be small compared to a characteristic length of the macroscopic fluid field L. We refer to the fluid phases by number $k = 1,2,3\ldots, M$, and to the solid phase by letter $k = w$. Assume that ψ_k is some quantity associated with phase k. We define the local volume average ${}^3\langle\psi_k\rangle$, intrinsic volume average ${}^{3i}\langle\psi_k\rangle$, and total local volume average $\langle\psi\rangle$, respectively, as follows:

$$ {}^3\langle\psi_k\rangle = \frac{1}{\upsilon}\int_{A_k} \psi_k \, d\upsilon \tag{2.5.1}$$

$$ {}^{3i}\langle\psi_k\rangle = \frac{1}{\upsilon_k}\int_{A_k} \psi_k \, d\upsilon \tag{2.5.2}$$

$$ \langle\psi\rangle = \sum_{k=1}^{M} {}^3\langle\psi_k\rangle + {}^3\langle\psi_w\rangle. \tag{2.5.3}$$

Here, ψ_k can be a scalar, vector, or tensor. We require with enveloping surface A to be sufficiently large that ${}^3\langle\psi_k\rangle$ is nearly independent of position over a distance of the same order of magnitude as the characteristic length ℓ (for all pertinent ψ and all phases k). This implies the following [1]:

$$^{3i}\langle{}^{3}\langle\psi_k\rangle\rangle = {}^{3j}\langle{}^{3}\langle\psi_k\rangle\rangle = \langle{}^{3}\langle\psi_k\rangle\rangle = {}^{3}\langle\psi_k\rangle \qquad (2.5.4)$$

$$^{3i}\langle{}^{3i}\langle\psi_k\rangle\rangle = {}^{3j}\langle{}^{3i}\langle\psi_k\rangle\rangle = \langle{}^{3i}\langle\psi_k\rangle\rangle = {}^{3i}\langle\psi_k\rangle \qquad (2.5.5)$$

$$^{3i}\langle\langle\psi\rangle\rangle = {}^{3j}\langle\langle\psi\rangle\rangle = \langle\langle\psi\rangle\rangle = \langle\psi\rangle. \qquad (2.5.6)$$

Although the local volume average ${}^{3}\langle\psi_k\rangle$ is assured of being a continuous function in space and time by its definition, its derivatives are generally only piecewise continuous. As we increase the characteristic length of ℓ with respect to the characteristic length of the local dispersed flow d, ${}^{3}\langle\psi_k\rangle$ approaches with arbitrarily small error a smooth, continuous function with continuous derivatives. But there is a limit to the maximum size of ℓ if its characteristic length is to be small compared with the characteristic length of the macroscopic flow L. In what follows, we assume as an approximation that ${}^{3}\langle\psi_k\rangle$ is a smooth continuous function of position and time with as many continuous derivatives as required.

It is understood that the size of characteristic length ℓ of local averaging volume is sufficiently large so that the relations as stated in Eqs. (2.5.4)–(2.5.6) are satisfied. The superscript j in Eqs. (2.5.4)–(2.5.6) denotes the fluid phase j.

3 Phasic conservation equations and interfacial balance equations

The equations of conservation for a pure phase are given by continuum mechanics. Although a pure phase commonly refers to one physical phase, such as vapor, liquid, or solid, it also includes certain nonreactive mixtures, such as room atmosphere or an aqueous solution of glycerine. The identification of a phase in a multiphase system is best made in terms of its dynamic phases according to their different dynamic responses, despite the fact that they may be of the same material. For example, within the framework of generalized multiphase mechanics first suggested by Soo [9], particles or bubbles of different ranges of sizes, densities, and shapes are treated as different dynamic phases.

3.1 Phasic conservation equations

For a pure phase k, the equations of continuity, momentum, and total energy are, respectively,

$$\left(\frac{\partial \rho_k}{\partial t}\right) + \nabla \cdot (\rho_k \underline{U}_k) = 0 \qquad (3.1.1)$$

$$\left(\frac{\partial \rho_k \underline{U}_k}{\partial t}\right) + \nabla \cdot (\rho_k \underline{U}_k \underline{U}_k) = -\nabla P_k + \nabla \cdot \underline{\underline{\tau}}_k + \rho_k \underline{f} \quad (3.1.2)$$

$$\left(\frac{\partial \rho_k E_k}{\partial t}\right) + \nabla \cdot (\rho_k \underline{U}_k E_k) = -\nabla \cdot \underline{U}_k P_k + \nabla \cdot (\underline{U}_k \cdot \underline{\underline{\tau}}_k) + \rho_k \underline{U}_k \cdot \underline{f} - \nabla \cdot \underline{J}_{qk} + J_{EK}, \quad (3.1.3)$$

where ρ_k is the density of fluid in pure phase k; $\underline{U}_k$ is its velocity; P_k is the static pressure; $\underline{f}$ is the field force per unit mass, which is taken to be a constant in the present study; $\underline{\underline{\tau}}_k$ is the viscous stress tensor; E_k is the total energy per unit mass; $\underline{J}_{qk}$ is the heat flux vector; and J_{Ek} is the heat source per unit volume inside phase k. By definition, $E_k = u_k + \underline{U}_k \cdot \underline{U}_k/2$, with u_k being the internal energy per unit mass. Alternatively, the energy equation may be expressed in terms of u_k (internal energy) or h_k (enthalpy) per unit mass:

$$\frac{\partial \rho_k u_k}{\partial t} + \nabla \cdot (\rho_k \underline{U}_k u_k) = -P_k \nabla \cdot \underline{U}_k - \nabla \cdot \underline{J}_{qk} + J_{Ek} + \underline{\underline{\tau}}_k : \nabla \underline{U}_k. \quad (3.1.4)$$

The double dot in the last term denotes the scalar product of two second-order tensors and is usually represented as Φ_k, the dissipation rate per unit volume of phase k,

$$\frac{\partial \rho_k h_k}{\partial t} + \nabla \cdot (\rho_k \underline{U}_k h_k) = \frac{dP_k}{dt_k} - \nabla \cdot \underline{J}_{qk} + J_{Ek} + \Phi_k, \quad (3.1.5)$$

in which the substantive derivative

$$\frac{d}{dt_k} = \frac{\partial}{\partial t} + \underline{U}_k \cdot \nabla. \quad (3.1.6)$$

3.2 Interfacial balance equations

The simplest case of the fluid–fluid interface is one of zero thickness [1]. The mass, momentum, and total energy balances at the interface A_{kf} (between phases k and f; Fig. 2.1) are given by the following:

Mass Balance:

$$\rho_k \left(\underline{U}_k - \underline{W}_k\right) \cdot \underline{n}_k + \rho_f(\underline{U}_f - \underline{W}_f) \cdot \underline{n}_f = 0 \qquad (3.2.1)$$

Momentum Balance (effect of changes in mean curvature can be ignored for diameters of bubbles and droplets greater than 0.1 mm [33]):

$$\begin{aligned} &-\nabla_{kf}\sigma_{kf} + 2\sigma_{kf}H_{kf}\underline{n}_k - \rho_k\underline{U}_k \left(\underline{U}_k - \underline{W}_k\right) \cdot \underline{n}_k \\ &\quad - \rho_f\underline{U}_f(\underline{U}_f - \underline{W}_f) \cdot \underline{n}_f + \left(-P_k\underline{\underline{I}} + \underline{\underline{\tau}}_k\right) \cdot \underline{n}_k \\ &\quad + \left(-P_f\underline{\underline{I}} + \underline{\underline{\tau}}_f\right) \cdot \underline{n}_f = 0 \end{aligned} \qquad (3.2.2)$$

Total Energy Balance (capillary energy ignored):

$$\begin{aligned} &\rho_k E_k \left(\underline{U}_k - \underline{W}_k\right) \cdot \underline{n}_k + \underline{J}_{qk} \cdot \underline{n}_k + \rho_f E_f(\underline{U}_f - \underline{W}_f) \cdot \underline{n}_f \\ &\quad + \underline{J}_{qf} \cdot \underline{n}_f - \underline{U}_k \cdot \left(-P_k\underline{\underline{I}} + \underline{\underline{\tau}}_k\right) \cdot \underline{n}_k \\ &\quad - \underline{U}_f \cdot \left(-P_f\underline{\underline{I}} + \underline{\underline{\tau}}_f\right) \cdot \underline{n}_f = 0, \end{aligned} \qquad (3.2.3)$$

where $\underline{n}_k$ is the unit normal vector outward from phase k and directed along the mean curvature H_{kf}, σ_{kf} is the interfacial tension, ∇_{kf} is the surface gradient operator, and $\underline{\underline{I}}$ is the unitary tensor. The interfacial velocity $\underline{W}_k = \underline{W}_f$ and H_{kf} is positive when the associated radius is pointing outward. In Eq. (3.2.3), the energy associated with surface tension and the corresponding dissipation are neglected.

The internal energy and enthalpy balance equations for the interface A_{kf} are

$$\rho_k u_k (\underline{U}_k - \underline{W}_k) \cdot \underline{n}_k + \underline{J}_{qk} \cdot \underline{n}_k + \rho_f u_f (\underline{U}_f - \underline{W}_f) \cdot \underline{n}_f + \underline{J}_{qf} \cdot \underline{n}_f = 0 \qquad (3.2.4)$$

and

$$\rho_k h_k (\underline{U}_k - \underline{W}_k) \cdot \underline{n}_k - P_k (\underline{U}_k - \underline{W}_k) \cdot \underline{n}_k + \underline{J}_{qk} \cdot \underline{n}_k + \rho_f h_f (\underline{U}_f - \underline{W}_f) \cdot \underline{n}_f - P_f (\underline{U}_f - \underline{W}_f) \cdot \underline{n}_f + \underline{J}_{qf} \cdot \underline{n}_f = 0. \qquad (3.2.5)$$

We note that only one of Eqs. (3.2.3) through (3.2.5) is independent. All variables in Eqs. (3.2.1) through (3.2.5), such as density, velocity, pressure, viscous stress, total energy, internal energy, and enthalpy, refer to interface A_{kf}.

In principle, the coupled phasic equations should be solved for given initial conditions together with boundary conditions at the phase interfaces. Because the configuration and location of the fluid–fluid interfaces are not generally known, however, their detailed solutions are next to impossible. When the length scale over which the point variables undergo significant changes is small compared with that over which the knowledge of these variables is of practical interest, information of their volume averages is all that is needed. A similar statement can be made regarding time-scale considerations. To preserve the identity of the dynamic phases, local volume averaging is performed first; this is done in Chapter 4. Time averaging of the volume-averaged equations is presented in Chapter 5.

4 Local volume-averaged conservation equations and interfacial balance equations

Application of the local volume-averaging theorems [Eqs. (2.4.1a), (2.4.1b), (2.4.2), and (2.4.7)] to the phasic conservation equations given in Chapter 3 leads to the following set of local volume-averaged conservation equations for multiphase flow. These equations are rigorous and subject only to the length-scale restriction, Eq. (2.4.3), which is inherent in the local volume-averaging theorems. Unless otherwise stated, all solid structures are stationary, nonporous, and nonreacting; $\underline{U}_k$ and $\underline{W}_k$ vanish in A_{wk}. Both volume porosities (γ_v) and directional surface porosities (γ_{Ax}, γ_{Ay}, and γ_{Az}) are invariant in time and in space, and they are functions of their initial structure locations, sizes, and shapes.

4.1 Local volume-averaged mass conservation equation of a phase and its interfacial balance equation

The continuity equation is written as

$$\left(\frac{\partial \rho_k}{\partial t}\right) + \nabla \cdot (\rho_k \underline{U}_k) = 0. \tag{3.1.1}$$

When the local volume-averaging theorems are applied to the terms in the continuity equation [Eq. (3.1.1)], Eqs. (2.2.2), (2.4.1b), (2.4.2), and (2.4.7) are employed to give

$$^{3}\left\langle \frac{\partial \rho_k}{\partial t} \right\rangle = \gamma_v \left(\frac{\partial \alpha_k \, {}^{3i}\langle \rho_k \rangle}{\partial t} \right) - v^{-1} \int_{A_k} \rho_k \underline{W}_k \cdot \underline{n}_k \, dA \quad (4.1.1)$$

and

$$\begin{aligned} ^{3}\langle \nabla \cdot (\rho_k \underline{U}_k) \rangle &= \nabla \cdot {}^{3}\langle \rho_k \underline{U}_k \rangle + v^{-1} \int_{A_k} \rho_k \underline{U}_k \cdot \underline{n}_k \, dA \\ &= \gamma_A \nabla \cdot \alpha_k \, {}^{2i}\langle \rho_k \underline{U}_k \rangle \\ &\quad + v^{-1} \int_{A_k} \rho_k \underline{U}_k \cdot \underline{n}_k \, dA \end{aligned} \quad (4.1.2)$$

because $A_k = A_{kf} + A_{wk}$ and $\underline{U}_k = 0$ on A_{wk}. Hence, the continuity equation takes the following form in terms of intrinsic average:

$$\begin{aligned} &\gamma_v \left(\frac{\partial \alpha_k \, {}^{3i}\langle \rho_k \rangle}{\partial t} \right) + \gamma_A \nabla \cdot \alpha_k \, {}^{2i}\langle \rho_k \underline{U}_k \rangle \\ &\quad = -v^{-1} \int_{A_k} \rho_k \left(\underline{U}_k - \underline{W}_k \right) \cdot \underline{n}_k \, dA \end{aligned} \quad (4.1.3)$$

or

$$\gamma_v \left(\frac{\partial \alpha_k \, {}^{3i}\langle \rho_k \rangle}{\partial t} \right) + \gamma_A \nabla \cdot \alpha_k \, {}^{2i}\langle \rho_k \underline{U}_k \rangle = \gamma_v \alpha_k \Gamma_k . \quad (4.1.4)$$

On the right-hand side (RHS) of Eqs. (4.1.3) and (4.1.4) is the net rate of interfacial mass generation of phase k over A_k per unit volume of v. Here, γ_A is the directional surface porosity, and γ_v is the volume porosity.

The local volume-averaged interfacial mass balance equation [Eq. (3.2.1)] now gives

$$v^{-1}\int_{A_k}\rho_k(\underline{U}_k-\underline{W}_k)\cdot\underline{n}_k\,dA=-v^{-1}\int_{A_f}\rho_f(\underline{U}_f-\underline{W}_f)\cdot\underline{n}_f\,dA$$
$$=-\gamma_v\alpha_k\Gamma_k=+\gamma_v\alpha_f\Gamma_f. \quad (4.1.5)$$

4.2 Local volume-averaged linear momentum equation and its interfacial balance equation

The linear momentum equation is written as follows:

$$\left(\frac{\partial\rho_k\underline{U}_k}{\partial t}\right)+\nabla\cdot(\rho_k\underline{U}_k\underline{U}_k)=-\nabla P_k+\nabla\cdot\underline{\underline{\tau}}_k+\rho_k f. \quad (3.1.2)$$

When the local volume-averaging theorems are applied to the terms of the linear momentum equation [Eq. (3.1.2)], Eqs. (2.2.2), (2.4.1a), (2.4.1b), (2.4.2), and (2.4.7) are employed to give

$$^3\left\langle\frac{\partial\rho_k\underline{U}_k}{\partial t}\right\rangle=\gamma_v\left\langle\frac{\partial\alpha_k\,{}^{3i}\langle\rho_k\underline{U}_k\rangle}{\partial t}\right\rangle-v^{-1}\int_{A_k}\rho_k\underline{U}_k\underline{W}_k\cdot\underline{n}_k\,dA$$
$$(4.2.1)$$

$$^3\langle\nabla\cdot\rho_k\underline{U}_k\underline{U}_k\rangle=\gamma_A\nabla\cdot\alpha_k\,{}^{2i}\langle\rho_k\underline{U}_k\underline{U}_k\rangle$$
$$+v^{-1}\int_{A_k}\rho_k\underline{U}_k\underline{U}_k\cdot\underline{n}_k\,dA \quad (4.2.2)$$

$$-{}^3\langle\nabla P_k\rangle=-\nabla^3\langle P_k\rangle-v^{-1}\int_{A_k}P_k\underline{n}_k\,dA$$
$$=-\gamma_v\nabla\alpha_k\,{}^{3i}\langle P_k\rangle-v^{-1}\int_{A_k}P_k\underline{n}_k\,dA \quad (4.2.3)$$

$$^3\langle\nabla\cdot\underline{\underline{\tau}}_k\rangle=\gamma_A\nabla\cdot\alpha_k\,{}^{2i}\langle\underline{\underline{\tau}}_k\rangle+v^{-1}\int_{A_k}\underline{\underline{\tau}}_k\cdot\underline{n}_k\,dA \quad (4.2.4)$$

and

$$^{3}\langle \rho_k \underline{f} \rangle = \gamma_v \alpha_k \, ^{3i}\langle \rho_k \rangle \underline{f}. \tag{4.2.5}$$

When the integrals of Eqs. (4.2.1)–(4.2.4) are combined, we note that $A_k = A_{kf} + A_{wk}$. The following terms can be grouped in two ways:

1. Combining integrals of Eqs. (4.2.3) and (4.2.4):

$$v^{-1}\int_{A_k} (-P_k \underline{\underline{I}} + \underline{\underline{\tau}}_k) \cdot \underline{n}_k \, dA = -v^{-1}\int_{A_{wk}} (-P_k \underline{\underline{I}} + \underline{\underline{\tau}}_k) \cdot \underline{n}_k \, dA$$
$$-v^{-1}\int_{A_{kf}} (-P_k \underline{\underline{I}} + \underline{\underline{\tau}}_k) \cdot \underline{n}_k \, dA. \tag{4.2.6}$$

   The first term on the RHS of Eq. (4.2.6) gives the distributed resistance force per unit volume of the phase k exerted by the stationary solid structures inside v. The second term on the RHS gives the interfacial drag and the diffusive force [31,32] on phase k by phase f in the mixture.

2. Combining integrals of Eqs. (4.2.1) and (4.2.2):

$$\begin{aligned} v^{-1}&\int_{A_k} \rho_k \underline{U}_k (\underline{U}_k - \underline{W}_k) \cdot \underline{n}_k \, dA \\ &= v^{-1}\int_{A_k} \rho_k (\underline{U}_k - \underline{W}_k)(\underline{U}_k - \underline{W}_k) \cdot \underline{n}_k \, dA \\ &\quad + v^{-1}\int_{A_k} \rho_k \underline{W}_k (\underline{U}_k - \underline{W}_k) \cdot \underline{n}_k \, dA \\ &= v^{-1}\int_{A_k} \rho_k (\underline{U}_k - \underline{W}_k)(\underline{U}_k - \underline{W}_k) \cdot \underline{n}_k \, dA \\ &\quad - \gamma_v \alpha_k \Gamma_k \overline{\underline{W}}_s, \end{aligned} \tag{4.2.7}$$

where $\underline{\overline{W}}_s$ is defined by

$$\underline{\overline{W}}_s = \frac{\upsilon^{-1} \int_{A_k} \underline{W}_k \rho_k \left(\underline{U}_k - \underline{W}_k\right) \cdot \underline{n}_k \, dA}{\upsilon^{-1} \int_{A_k} \rho_k \left(\underline{U}_k - \underline{W}_k\right) \cdot \underline{n}_k \, dA}, \quad (4.2.8)$$

which is a mean interface velocity at which phase k is generated or removed over A_{kf}.

Note that the momentum transfer integral in Eq. (4.2.7) does not necessarily go to zero when $\Gamma_k = 0$ [Eq. (4.1.4)] because of the condition given in Eq. (2.4.3), and the local averaging volume υ should not be made arbitrarily small. It is conceivable that condensation may occur over a portion of A_{kf} and evaporation over the remaining portion. Thus, the integral may vanish (i.e., $\Gamma_k = 0$). It does not necessarily follow that, locally, $\left(\underline{U}_k - \underline{W}_k\right) \cdot \underline{n}_k = 0$.

The local volume-averaged momentum equation takes the following form in terms of intrinsic averages by combining Eqs. (4.2.1)–(4.2.7).

$$\begin{aligned}
&\gamma_\upsilon \frac{\partial}{\partial t} \left(\alpha_k{}^{3i} \langle \rho_k \underline{U}_k \rangle\right) + \gamma_A \nabla \cdot \alpha_k{}^{2i} \langle \rho_k \underline{U}_k \underline{U}_k \rangle \\
&= -\gamma_\upsilon \nabla \alpha_k{}^{3i} \langle P_k \rangle + \gamma_A \nabla \cdot \alpha_k{}^{2i} \langle \underline{\underline{\tau}}_k \rangle + \gamma_\upsilon \alpha_k{}^{3i} \langle \rho_k \rangle \underline{f} \\
&\quad + \upsilon^{-1} \int_{A_k} \left(-P_k \underline{\underline{I}} + \underline{\underline{\tau}}_k\right) \cdot \underline{n}_k \, dA \\
&\quad - \upsilon^{-1} \int_{A_k} \rho_k \underline{U}_k \left(\underline{U}_k - \underline{W}_k\right) \cdot \underline{n}_k \, dA. \qquad (4.2.9)
\end{aligned}$$

The last term in Eq. (4.2.9) may be replaced by its equivalent, Eq. (4.2.7).

For a pure phase k, $\alpha_k = 1$, $A_k = 0$, without any structure, $\gamma_{\upsilon} = \gamma_{A} = 1$. In the absence of any motion, $\underline{\underline{\tau}}_k$ is a function of $\underline{U}_k$; any quantities associated with $\underline{U}_k$ are zero. Equation (4.2.9) gives

$$\nabla\, {}^{3i}\langle P_k\rangle - {}^{3i}\langle \rho_k\rangle \underline{f} = 0 \tag{4.2.10}$$

or

$$\nabla P_k = \rho_k \underline{g}, \tag{4.2.11}$$

with $\underline{f} = \underline{g}$, thus satisfying the basic relation of hydrostatics.

For bubbles and droplets, the first two terms in Eq. (3.2.2) give rise to capillary pressure difference across the interface. Consequently, we define an average capillary pressure difference as follows:

$$\upsilon^{-1}\int_{A_k} \Big(\langle P_{ck}\rangle - \langle P_{cf}\rangle\Big)\underline{n}_k\, dA$$
$$= \upsilon^{-1}\int_{A_k} \Big(-\nabla_{kf}\sigma_{kf} + 2\sigma_{kf}H_{kf}\underline{n}_k\Big)\, dA. \tag{4.2.12}$$

The local volume-averaged interfacial linear momentum balance equation is, from Eqs. (3.2.2) and (4.2.12),

$$\upsilon^{-1}\int_{A_k} \rho_k \underline{U}_k\,(\underline{U}_k - \underline{W}_k)\cdot \underline{n}_k\, dA$$
$$+\,\upsilon^{-1}\int_{A_k} \big[(P_k - \langle P_{ck}\rangle)\underline{\underline{I}} - \underline{\underline{\tau}}_k\big]\cdot \underline{n}_k\, dA$$
$$= -\upsilon^{-1}\int_{A_f} \rho_f \underline{U}_f(\underline{U}_f - \underline{W}_f)\cdot \underline{n}_f\, dA$$
$$-\,\upsilon^{-1}\int_{A_f} \big[(P_f - \langle P_{cf}\rangle)\underline{\underline{I}} - \underline{\underline{\tau}}_f\big]\cdot \underline{n}_f\, dA. \tag{4.2.13}$$

Generally speaking, the capillary pressure difference can be neglected for bubbles or droplets of diameters greater than 0.1 mm [33].

4.3 Local volume-averaged total energy equation and its interfacial balance equation

The total energy equation is written as follows:

$$\left(\frac{\partial \rho_k E_k}{\partial t}\right) + \nabla \cdot (\rho_k \underline{U}_k E_k) = -\nabla \cdot \underline{U}_k P_k + \nabla \cdot (\underline{U}_k \cdot \underline{\underline{\tau}}_k) + \rho_k \underline{U}_k \cdot \underline{f} - \nabla \cdot \underline{J}_{qk} + J_{EK}. \tag{3.1.3}$$

When the local volume-averaging theorems are applied to the terms of the total energy equation [Eq. (3.1.3)], Eqs. (2.2.2), (2.4.1b), (2.4.2), and (2.4.7) are employed to give

$$^{3}\left\langle\frac{\partial \rho_k E_k}{\partial t}\right\rangle = \gamma_\upsilon \frac{\partial \alpha_k {}^{3i}\langle \rho_k E_k\rangle}{\partial t} - \upsilon^{-1}\int_{A_k} \rho_k E_k \underline{W}_k \cdot \underline{n}_k \, dA \tag{4.3.1}$$

$$^{3}\langle \nabla \cdot (\rho_k \underline{U}_k E_k)\rangle = \gamma_A \nabla \cdot \alpha_k {}^{2i}\langle \rho_k \underline{U}_k E_k\rangle + \upsilon^{-1}\int_{A_k} \rho_k \underline{U}_k E_k \cdot \underline{n}_k \, dA \tag{4.3.2}$$

$$^{3}\langle \nabla \cdot (\underline{U}_k P_k)\rangle = \gamma_A \nabla \cdot \alpha_k {}^{2i}\langle \underline{U}_k P_k\rangle + \upsilon^{-1}\int_{A_k} \underline{U}_k P_k \cdot \underline{n}_k \, dA \tag{4.3.3}$$

$$^{3}\langle \nabla \cdot (\underline{U}_k \cdot \underline{\underline{\tau}}_k)\rangle = \gamma_A \nabla \cdot \alpha_k {}^{2i}\langle \underline{U}_k \cdot \underline{\underline{\tau}}_k\rangle + \upsilon^{-1}\int_{A_k} \underline{U}_k \cdot \underline{\underline{\tau}}_k \cdot \underline{n}_k \, dA \tag{4.3.4}$$

$$ {}^{3}\langle \rho_k \underline{U}_k \rangle \cdot \underline{f} = \gamma_v \alpha_k \, {}^{3i}\langle \rho_k \underline{U}_k \rangle \cdot \underline{f} \tag{4.3.5} $$

$$ \begin{aligned} {}^{3}\langle \nabla \cdot \underline{J}_{qk} \rangle &= \gamma_A \nabla \cdot \alpha_k \, {}^{2i}\langle \underline{J}_{qk} \rangle + v^{-1} \int_{A_{kf}} \underline{J}_{qk} \cdot \underline{n}_k \, dA \\ &\quad + v^{-1} \int_{A_{wk}} \underline{J}_{qk} \cdot \underline{n}_k \, dA \\ &= \gamma_A \nabla \cdot \alpha_k \, {}^{2i}\langle \underline{J}_{qk} \rangle - \gamma_v \alpha_k \, {}^{3i}\langle \dot{Q}_{kf} \rangle \\ &\quad - \gamma_v \alpha_k \, {}^{3i}\langle \dot{Q}_{wk} \rangle \end{aligned} \tag{4.3.6} $$

and

$$ {}^{3}\langle J_{Ek} \rangle = \gamma_v \alpha_k \, {}^{3i}\langle J_{Ek} \rangle \tag{4.3.7} $$

with

$$ \gamma_v \alpha_k \, {}^{3i}\langle \dot{Q}_{kf} \rangle = -v^{-1} \int_{A_{kf}} \underline{J}_{qk} \cdot \underline{n}_k \, dA \tag{4.3.8} $$

and

$$ \gamma_v \alpha_k \, {}^{3i}\langle \dot{Q}_{wk} \rangle = -v^{-1} \int_{A_{wk}} \underline{J}_{qk} \cdot \underline{n}_k \, dA, \tag{4.3.9} $$

where ${}^{3i}\langle \dot{Q}_{kf} \rangle$ is the rate of fluid–fluid interfacial heat transport into phase k per unit volume of phase k, and ${}^{3i}\langle \dot{Q}_{wk} \rangle$ is the rate of fluid–solid interfacial heat transport into phase k per unit volume of phase k. Here, ${}^{3i}\langle \dot{Q}_{kf} \rangle$ would depend on the physical and chemical processes occurring in A_{kf} as well as its configuration.

And, again,

$$ \begin{aligned} & v^{-1} \int_{A_k} \rho_k E_k \left(\underline{U}_k - \underline{W}_k \right) \cdot \underline{n}_k \, dA \\ &= v^{-1} \int_{A_k} \rho_k \left(E_k - E_{kf} \right) \left(\underline{U}_k - \underline{W}_k \right) \cdot \underline{n}_k \, dA \\ &\quad - \gamma_v \alpha_k \Gamma_k \overline{E}_s, \end{aligned} \tag{4.3.10} $$

where $\overline{E}_s$ is the mean interface total energy given by

$$\overline{E}_s = \frac{v^{-1}\int_{A_k} \rho_k E_{kf}\left(\underline{U}_k - \underline{W}_k\right)\cdot \underline{n}_k\, dA}{v^{-1}\int_{A_k} \rho_k\left(\underline{U}_k - \underline{W}_k\right)\cdot \underline{n}_k\, dA}, \qquad (4.3.11)$$

which is a mean interface total energy associated with generation of phase k over A_{kf}.

The local volume-averaged total energy equation takes the form in terms of intrinsic averages by combining Eqs. (4.3.1)–(4.3.9):

$$\begin{aligned}
&\gamma_v \frac{\partial}{\partial t}\left(\alpha_k\, {}^{3i}\langle\rho_k E_k\rangle\right) + \gamma_A \nabla\cdot\alpha_k\, {}^{2i}\langle\rho_k \underline{U}_k E_k\rangle \\
&= -\gamma_A \nabla\cdot\alpha_k\, {}^{2i}\langle\underline{U}_k P_k\rangle + \gamma_A \nabla\cdot\alpha_k\, {}^{2i}\langle\underline{U}_k\cdot\underline{\underline{\tau}}_k\rangle - \gamma_A \nabla\cdot\alpha_k\, {}^{2i}\langle\underline{J}_{qk}\rangle \\
&\quad + \gamma_v \alpha_k\left({}^{3i}\langle\rho_k \underline{U}_k\rangle\cdot\underline{f} + {}^{3i}\langle J_{Ek}\rangle + {}^{3i}\langle\dot{Q}_{kf}\rangle + {}^{3i}\langle\dot{Q}_{wk}\rangle\right) \\
&\quad + v^{-1}\int_{A_k}\left(-P_k\underline{U}_k + \underline{\underline{\tau}}_k\cdot\underline{U}_k\right)\cdot\underline{n}_k\, dA \\
&\quad - v^{-1}\int_{A_k}\rho_k E_k\left(\underline{U}_k - \underline{W}_k\right)\cdot\underline{n}_k\, dA. \qquad (4.3.12)
\end{aligned}$$

The last term on the RHS of Eq. (4.3.12) may be replaced by its equivalent, Eq. (4.3.10).

Ignoring capillary energy, the local volume-averaged total energy balance equation for the interface A_{kf} is, from Eq. (3.2.3),

$$\begin{aligned}
&- v^{-1}\int_{A_k}\rho_k E_k\left(\underline{U}_k - \underline{W}_k\right)\cdot\underline{n}_k\, dA - v^{-1}\int_{A_k}\underline{J}_{qk}\cdot\underline{n}_k\, dA \\
&+ v^{-1}\int_{A_k}\underline{U}_k\cdot\left(-P_k\underline{\underline{I}} + \underline{\underline{\tau}}_k\right)\cdot\underline{n}_k\, dA \\
&- v^{-1}\int_{A_f}\rho_f E_f\left(\underline{U}_f - \underline{W}_f\right)\cdot\underline{n}_k\, dA - v^{-1}\int_{A_f}\underline{J}_{qf}\cdot\underline{n}_f\, dA \\
&+ v^{-1}\int_{A_f}\underline{U}_f\cdot\left(-P_f\underline{\underline{I}} + \underline{\underline{\tau}}_f\right)\cdot\underline{n}_f\, dA = 0. \qquad (4.3.13)
\end{aligned}$$

4.4 Local volume-averaged internal energy equation and its interfacial balance equation

Alternatively, the energy equation can be expressed in terms of internal energy instead of total energy as shown in Eq. (3.1.4). For phase k, we have the following:

$$\frac{\partial \rho_k u_k}{\partial t} + \nabla \cdot (\rho_k \underline{U}_k u_k) = -P_k \nabla \cdot \underline{U}_k - \nabla \cdot \underline{J}_{qk} + J_{Ek} + \underline{\underline{\tau}}_k : \nabla \underline{U}_k. \tag{3.1.4}$$

The double dot in the last term denotes the scalar product of two second-order tensors and is usually represented as Φ_k, the dissipation rate per unit volume of phase k.

When the local volume-averaging theorems are applied to the terms in the internal energy equation [Eq. (3.1.4)], Eqs. (2.2.2), (2.4.1b), (2.4.2), and (2.4.7) are employed to give the following:

$$^3\left\langle \frac{\partial \rho_k u_k}{\partial t} \right\rangle = \gamma_v \frac{\partial \alpha_k\, ^{3i}\langle \rho_k u_k \rangle}{\partial t} - v^{-1} \int_{A_k} \rho_k u_k \underline{W}_k \cdot \underline{n}_k dA \tag{4.4.1}$$

$$\begin{aligned} ^3\langle \nabla \cdot (\rho_k \underline{U}_k u_k) \rangle &= \nabla \cdot {}^3\langle \rho_k \underline{U}_k u_k \rangle + v^{-1} \int_{A_k} \rho_k \underline{U}_k u_k \cdot \underline{n}_k dA \\ &= \gamma_A \nabla \cdot \alpha_k\, ^{2i}\langle \rho_k \underline{U}_k u_k \rangle \\ &\quad + v^{-1} \int_{A_k} \rho_k \underline{U}_k u_k \cdot \underline{n}_k dA \end{aligned} \tag{4.4.2}$$

$$^3\langle P_k \nabla \cdot \underline{U}_k \rangle = \gamma_v \alpha_k\, ^{3i}\langle P_k \nabla \cdot \underline{U}_k \rangle \tag{4.4.3}$$

$$\begin{aligned} ^3\langle \nabla \cdot \underline{J}_{qk} \rangle &= \gamma_A \nabla \cdot \alpha_k\, ^{2i}\langle \underline{J}_{qk} \rangle + v^{-1} \int_{A_{kf}} \underline{J}_{qk} \cdot \underline{n}_k dA \\ &\quad + v^{-1} \int_{A_{wk}} \underline{J}_{qk} \cdot \underline{n}_k dA \\ &= \gamma_A \nabla \cdot \alpha_k\, ^{2i}\langle \underline{J}_{qk} \rangle - \gamma_v \alpha_k\, ^{3i}\langle \dot{Q}_{kf} \rangle - \gamma_v \alpha_k\, ^{3i}\langle \dot{Q}_{wk} \rangle \end{aligned} \tag{4.3.6}$$

$$^{3}\langle\Phi_k\rangle = \gamma_v \alpha_k {}^{3i}\langle\Phi_k\rangle \tag{4.4.4}$$

$$^{3}\langle J_{Ek}\rangle = \gamma_v \alpha_k {}^{3i}\langle J_{Ek}\rangle. \tag{4.3.7}$$

The local volume-averaged internal energy equation is expressed in terms of intrinsic averages by combining Eqs. (4.4.1)–(4.4.3), (4.3.6), (4.4.4), and (4.3.7):

$$\begin{aligned}
&\gamma_v \frac{\partial}{\partial t}\left(\alpha_k {}^{3i}\langle\rho_k u_k\rangle\right) + \gamma_A \nabla \cdot \alpha_k {}^{2i}\langle\rho_k \underline{U}_k u_k\rangle \\
&= -\gamma_v \alpha_k {}^{3i}\langle P_k \nabla \cdot \underline{U}_k\rangle - \gamma_A \nabla \cdot \alpha_k {}^{2i}\langle \underline{J}_{qk}\rangle \\
&\quad + \gamma_v \alpha_k \left({}^{3i}\langle\Phi_k\rangle + {}^{3i}\langle J_{Ek}\rangle + {}^{3i}\langle\dot{Q}_{kf}\rangle + {}^{3i}\langle\dot{Q}_{wk}\rangle\right) \\
&\quad - v^{-1}\int_{A_k} \rho_k u_k \left(\underline{U}_k - \underline{W}_k\right) \cdot \underline{n}_k dA.
\end{aligned} \tag{4.4.5}$$

Again, we note

$$\begin{aligned}
&v^{-1}\int_{A_k} \rho_k u_k \left(\underline{U}_k - \underline{W}_k\right) \cdot \underline{n}_k dA \\
&= v^{-1}\int_{A_k} \rho_k \left(u_k - u_{kf}\right)\left(\underline{U}_k - \underline{W}_k\right) \cdot \underline{n}_k dA \\
&\quad - \gamma_v \alpha_k \Gamma_k \overline{u}_s,
\end{aligned} \tag{4.4.6}$$

where

$$\overline{u}_s = \frac{v^{-1}\int_{A_k} \rho_k u_{kf}\left(\underline{U}_k - \underline{W}_k\right) \cdot \underline{n}_k dA}{v^{-1}\int_{A_k} \rho_k \left(\underline{U}_k - \underline{W}_k\right) \cdot \underline{n}_k dA}, \tag{4.4.7}$$

which is the mean interface internal energy associated with generation of phase k over A_{kf}.

Also,

$$-v^{-1}\int_{A_{kf}} \underline{J}_{qk} \cdot \underline{n}_k dA = \gamma_v \alpha_k {}^{3i}\langle\dot{Q}_{kf}\rangle. \tag{4.3.8}$$

The last term on the RHS of Eq. (4.4.5) may be replaced by its equivalent, Eq. (4.4.6). The local volume-averaged internal energy balance equation for the interface A_{kf} is, from Eq. (3.2.4),

$$\begin{aligned} &\upsilon^{-1}\int_{A_k} \rho_k u_k \left(\underline{U}_k - \underline{W}_k\right)\cdot \underline{n}_k dA + \upsilon^{-1}\int_{A_k} \underline{J}_{qk}\cdot \underline{n}_k dA \\ &+ \upsilon^{-1}\int_{A_f} \rho_f u_f (\underline{U}_f - \underline{W}_f)\cdot \underline{n}_f dA \\ &+ \upsilon^{-1}\int_{A_f} \underline{J}_{qf}\cdot \underline{n}_f dA = 0. \end{aligned} \tag{4.4.8}$$

4.5 Local volume-averaged enthalpy equation and its interfacial balance equation

For phase k, we have the enthalpy equation

$$\frac{\partial\left(\rho_k h_k\right)}{\partial t} + \nabla\cdot\left(\rho_k \underline{U}_k h_k\right) = \frac{dP_k}{dt_k} - \nabla\cdot \underline{J}_{qk} + J_{Ek} + \Phi_k, \tag{3.1.5}$$

which is valid for an ideal gas. For constant ρ_k, the dP_k/dt_k term should be deleted regardless of whether the flow is steady or unsteady. Otherwise,

$$\frac{dP_k}{dt_k} = \frac{\partial P_k}{\partial t} + \underline{U}_k\cdot\nabla P_k = \frac{\partial P_k}{\partial t} + \nabla\cdot\left(P_k \underline{U}_k\right) - P_k \nabla\cdot \underline{U}_k. \tag{4.5.1}$$

By combining Eqs. (3.1.5) and (4.5.1), we obtain the enthalpy equations as follows:

$$\begin{aligned} \frac{\partial\left(\rho_k h_k\right)}{\partial t} + \nabla\cdot\left(\rho_k \underline{U}_k h_k\right) = {}& \frac{\partial P_k}{\partial t} + \nabla\cdot\left(P_k \underline{U}_k\right) - P_k \nabla\cdot \underline{U}_k \\ & - \nabla\cdot \underline{J}_{qk} + J_{Ek} + \Phi_k. \end{aligned} \tag{4.5.2}$$

When the local volume-averaging theorems are applied to the terms in the enthalpy equation [Eq. (4.5.2)], Eqs. (2.2.2), (2.4.1b), (2.4.2), and (2.4.7) are employed to give the following:

$$
\begin{aligned}
{}^3\left\langle \frac{\partial (\rho_k h_k)}{\partial t} \right\rangle &= \frac{\partial}{\partial t}{}^3\langle \rho_k h_k \rangle - \upsilon^{-1} \int_{A_k} \rho_k h_k \underline{W}_k \cdot \underline{n}_k \, dA \\
&= \gamma_\upsilon \frac{\partial}{\partial t} \alpha_k \, {}^{3i}\langle \rho_k h_k \rangle - \upsilon^{-1} \int_{A_k} \rho_k h_k \underline{W}_k \cdot \underline{n}_k \, dA
\end{aligned}
\tag{4.5.3}
$$

$$
\begin{aligned}
{}^3\langle \nabla \cdot \rho_k \underline{U}_k h_k \rangle &= \nabla \cdot {}^3\langle \rho_k \underline{U}_k h_k \rangle + \upsilon^{-1} \int_{A_k} \rho_k \underline{U}_k h_k \cdot \underline{n}_k \, dA \\
&= \gamma_A \nabla \cdot \alpha_k \, {}^{2i}\langle \rho_k \underline{U}_k h_k \rangle \\
&\quad + \upsilon^{-1} \int_{A_k} \rho_k \underline{U}_k h_k \cdot \underline{n}_k \, dA
\end{aligned}
\tag{4.5.4}
$$

$$
\begin{aligned}
{}^3\left\langle \frac{\partial P_k}{\partial t} \right\rangle &= \frac{\partial}{\partial t}{}^3\langle P_k \rangle - \upsilon^{-1} \int_{A_k} P_k \underline{W}_k \cdot \underline{n}_k \, dA \\
&= \gamma_\upsilon \frac{\partial}{\partial t} \alpha_k \, {}^{3i}\langle P_k \rangle - \upsilon^{-1} \int_{A_k} P_k \underline{W}_k \cdot \underline{n}_k \, dA
\end{aligned}
\tag{4.5.5}
$$

$$
\begin{aligned}
{}^3\langle \nabla \cdot (P_k \underline{U}_k) \rangle &= \nabla \cdot {}^3\langle P_k \underline{U}_k \rangle + \upsilon^{-1} \int_{A_k} P_k \underline{U}_k \cdot \underline{n}_k \, dA \\
&= \gamma_A \nabla \cdot \alpha_k \, {}^{2i}\langle P_k \underline{U}_k \rangle + \upsilon^{-1} \int_{A_k} P_k \underline{U}_k \cdot \underline{n}_k \, dA
\end{aligned}
\tag{4.5.6}
$$

$$
{}^3\langle P_k \nabla \cdot \underline{U}_k \rangle = \gamma_\upsilon \alpha_k \, {}^{3i}\langle P_k \nabla \cdot \underline{U}_k \rangle \tag{4.4.3}
$$

$$
{}^3\langle \nabla \cdot \underline{J}_{qk} \rangle = \gamma_A \nabla \cdot \alpha_k \, {}^{2i}\langle \underline{J}_{qk} \rangle - \gamma_\upsilon \alpha_k \, {}^{3i}\langle \dot{Q}_{kf} \rangle - \gamma_\upsilon \alpha_k \, {}^{3i}\langle \dot{Q}_{wk} \rangle \tag{4.3.6}
$$

$$^{3}\langle \Phi_k \rangle = \gamma_v \alpha_k \, ^{3i}\langle \Phi_k \rangle \tag{4.4.4}$$

$$^{3}\langle J_{Ek} \rangle = \gamma_v \alpha_k \, ^{3i}\langle J_{Ek} \rangle. \tag{4.3.7}$$

The local volume-averaged enthalpy equation in terms of intrinsic averages becomes

$$\begin{aligned}
&\gamma_v \frac{\partial}{\partial t}\left(\alpha_k \, ^{3i}\langle \rho_k h_k \rangle\right) + \gamma_A \nabla \cdot \alpha_k \, ^{2i}\langle \rho_k \underline{U}_k h_k \rangle \\
&= \gamma_v \frac{\partial}{\partial t}\left(\alpha_k \, ^{3i}\langle P_k \rangle\right) + \gamma_A \nabla \cdot \alpha_k \, ^{2i}\langle P_k \underline{U}_k \rangle - \gamma_v \alpha_k \, ^{3i}\langle P_k \nabla \cdot \underline{U}_k \rangle \\
&\quad - \gamma_A \nabla \cdot \alpha_k \, ^{2i}\langle \underline{J}_{qk} \rangle + \gamma_v \alpha_k \left(^{3i}\langle \Phi_k \rangle + \, ^{3i}\langle J_{Ek} \rangle \right. \\
&\quad \left. + \, ^{3i}\langle \dot{Q}_{kf} \rangle + \, ^{3i}\langle \dot{Q}_{wk} \rangle \right) \\
&\quad + \upsilon^{-1} \int_{A_k} P_k \left(\underline{U}_k - \underline{W}_k\right) \cdot \underline{n}_k dA \\
&\quad - \upsilon^{-1} \int_{A_k} \rho_k h_k \left(\underline{U}_k - \underline{W}_k\right) \cdot \underline{n}_k dA.
\end{aligned} \tag{4.5.7}$$

We again note

$$\begin{aligned}
&\upsilon^{-1} \int_{A_k} \rho_k h_k \left(\underline{U}_k - \underline{W}_k\right) \cdot \underline{n}_k dA \\
&= \upsilon^{-1} \int_{A_k} \rho_k \left(h_k - h_{kf}\right) \left(\underline{U}_k - \underline{W}_k\right) \cdot \underline{n}_k dA \\
&\quad - \gamma_v \alpha_k \Gamma_k \overline{h}_s,
\end{aligned} \tag{4.5.8}$$

where

$$\overline{h}_s = \frac{\upsilon^{-1} \int_{A_k} \rho_k h_{kf} \left(\underline{U}_k - \underline{W}_k\right) \cdot \underline{n}_k dA}{\upsilon^{-1} \int_{A_k} \rho_k \left(\underline{U}_k - \underline{W}_k\right) \cdot \underline{n}_k dA}, \tag{4.5.9}$$

which is the mean interface enthalpy associated with the generation of phase k over A_k. The last term on RHS of Eq. (4.5.7) may be replaced by its equivalent, Eq. (4.5.8).

The local volume-averaged enthalpy balance equation for interface A_{kf} is, from Eq. (3.2.5),

$$\upsilon^{-1}\int_{A_k} P_k\left(\underline{U}_k-\underline{W}_k\right)\cdot\underline{n}_k dA-\upsilon^{-1}\int_{A_k}\rho_k h_k\left(\underline{U}_k-\underline{W}_k\right)\cdot\underline{n}_k dA$$
$$-\upsilon^{-1}\int_{A_k}\underline{J}_{qk}\cdot\underline{n}_k dA$$
$$=-\upsilon^{-1}\int_{A_f}P_f(\underline{U}_f-\underline{W}_f)\cdot\underline{n}_f dA+\upsilon^{-1}\int_{A_f}\rho_f h_f(\underline{U}_f-\underline{W}_f)$$
$$\cdot\underline{n}_f dA+\upsilon^{-1}\int_{A_f}\underline{J}_{qf}\cdot\underline{n}_f dA. \qquad (4.5.10)$$

4.6 Summary of local volume-averaged conservation equations

In previous papers [2–6], the local volume-averaged conservation equations for multiphase flow in the presence of stationary and solid structures were presented. For the convenience of the reader, they are listed in this section.

4.6.1 Local volume-averaged mass conservation equation

$$\gamma_\upsilon\frac{\partial}{\partial t}\alpha_k{}^{3i}\langle\rho_k\rangle+\gamma_A\nabla\cdot\alpha_k{}^{2i}\langle\rho_k\underline{U}_k\rangle$$
$$=-\upsilon^{-1}\int_{A_k}\rho_k\left(\underline{U}_k-\underline{W}_k\right)\cdot\underline{n}_k dA, \qquad (4.6.1)$$

where A_k is the total fluid interfacial area associated with phase k inside υ. For the area integrals, $\underline{U}_k$ vanishes on A_{wk} for stationary and solid structures inside υ. The term on the

RHS of Eq. (4.6.1) denotes the rate of total interfacial mass generation of phase k per unit volume of υ. Denoting it by $\gamma_\upsilon \alpha_k \Gamma_k$, we have

$$\gamma_\upsilon \alpha_k \Gamma_k = -\upsilon^{-1} \int_{A_k} \rho_k \left(\underline{U}_k - \underline{W}_k\right) \cdot \underline{n}_k dA. \quad (4.6.1a)$$

4.6.2 Local volume-averaged linear momentum conservation equation

$$\begin{aligned}
&\gamma_\upsilon \frac{\partial}{\partial t} \left(\alpha_k {}^{3i}\langle \rho_k \underline{U}_k \rangle\right) + \gamma_A \nabla \cdot \alpha_k {}^{2i}\langle \rho_k \underline{U}_k \underline{U}_k \rangle \\
&= -\gamma_\upsilon \nabla \alpha_k {}^{3i}\langle P_k \rangle + \gamma_A \nabla \cdot \alpha_k {}^{2i}\langle \underline{\underline{\tau}}_k \rangle + \gamma_\upsilon \alpha_k {}^{3i}\langle \rho_k \rangle \underline{f} \\
&\quad + \upsilon^{-1} \int_{A_k} \left(-P_k \underline{\underline{I}} + \underline{\underline{\tau}}_k\right) \cdot \underline{n}_k dA \\
&\quad - \upsilon^{-1} \int_{A_k} \rho_k \underline{U}_k (\underline{U}_k - \underline{W}_k) \cdot \underline{n}_k dA, \qquad (4.6.2)
\end{aligned}$$

in which the field force per unit mass $\underline{f}$ is taken to be constant. As we noted previously, $A_k = A_{wk} + A_{kf}$. Thus, the fourth term on the RHS of Eq. (4.6.2) can be written as

$$\begin{aligned}
&\upsilon^{-1} \int_{A_k} (-P_k \underline{\underline{I}} + \underline{\underline{\tau}}_k) \cdot \underline{n}_k dA \\
&= \upsilon^{-1} \int_{A_{wk}} (-P_k \underline{\underline{I}} + \underline{\underline{\tau}}_k) \cdot \underline{n}_k dA \\
&\quad + \upsilon^{-1} \int_{A_{kf}} (-P_k \underline{\underline{I}} - \underline{\underline{\tau}}_k) \cdot \underline{n}_k dA \\
&= -\gamma_\upsilon \alpha_k {}^{3i}\langle \underline{R}_k \rangle + \upsilon^{-1} \int_{A_{kf}} (-P_k \underline{\underline{I}} - \underline{\underline{\tau}}_k) \cdot \underline{n}_k dA, \qquad (4.6.2a)
\end{aligned}$$

where $\gamma_\upsilon \alpha_k {}^{3i}\langle \underline{R}_k \rangle$ is the distributed flow resistance or distributed resistance per unit volume of phase k exerted by the fixed solids inside υ_k. The alternative local volume-averaged linear momentum conservative equation can be written as

$$\begin{aligned}
&\gamma_v \frac{\partial}{\partial t}(\alpha_k \, {}^{3i}\langle \rho_k \underline{U}_k \rangle) + \gamma_A \nabla \cdot \alpha_k \, {}^{2i}\langle \rho_k \underline{U}_k \underline{U}_k \rangle \\
&= -\gamma_v \nabla \alpha_k \, {}^{3i}\langle P_k \rangle + \gamma_A \nabla \cdot \alpha_k \, {}^{2i}\langle \underline{\underline{\tau}}_k \rangle + \gamma_v \alpha_k ({}^{3i}\langle \rho_k \rangle \underline{f}_k - {}^{3i}\langle \underline{R}_k \rangle) \\
&\quad + v^{-1} \int_{A_{kf}} (-P_k \underline{\underline{I}} + \underline{\underline{\tau}}_k) \cdot \underline{n}_k \, dA \\
&\quad - v^{-1} \int_{A_k} \rho_k \underline{U}_k (\underline{U}_k - \underline{W}_k) \cdot \underline{n}_k dA. \qquad (4.6.2b)
\end{aligned}$$

4.6.3 Local volume-averaged energy conservation equations

The local volume-averaged energy conservation equations are listed in this section.

4.6.3.1 In terms of total energy E_k, $E_k = u_k + \frac{1}{2}\underline{U}_k \cdot \underline{U}_k$

$$\begin{aligned}
&\gamma_v \frac{\partial}{\partial t} \alpha_k \, {}^{3i}\langle \rho_k E_k \rangle + \gamma_A \nabla \cdot \alpha_k \, {}^{2i}\langle \rho_k \underline{U}_k E_k \rangle \\
&= -\gamma_A \nabla \cdot \alpha_k \, {}^{2i}\langle \underline{U}_k P_k \rangle + \gamma_A \nabla \cdot \alpha_k \, {}^{2i}\langle \underline{U}_k \cdot \underline{\underline{\tau}}_k \rangle \\
&\quad - \gamma_A \nabla \cdot \alpha_k \, {}^{2i}\langle \underline{J}_{qk} \rangle \\
&\quad + \gamma_v \alpha_k ({}^{3i}\langle \rho_k \underline{U}_k \rangle \cdot \underline{f} + {}^{3i}\langle J_{Ek} \rangle + {}^{3i}\langle \dot{Q}_{kf} \rangle + {}^{3i}\langle \dot{Q}_{wk} \rangle) \\
&\quad + v^{-1} \int_{A_k} (-P_k \underline{U}_k + \underline{\underline{\tau}}_k \cdot \underline{U}_k) \cdot \underline{n}_k dA \\
&\quad - v^{-1} \int_{A_k} \rho_k E_k (\underline{U}_k - \underline{W}_k) \cdot \underline{n}_k dA, \qquad (4.6.3)
\end{aligned}$$

where ${}^{3i}\langle \dot{Q}_{kf} \rangle$ denotes the rate of fluid–fluid interfacial heat transfer into phase k per unit volume of phase k. It is defined by

$$\gamma_v \alpha_k \, {}^{3i}\langle \dot{Q}_{kf} \rangle = -v^{-1} \int_{A_{kf}} \underline{J}_{qk} \cdot \underline{n}_k dA. \qquad (4.3.8)$$

Here, ${}^{3i}\langle \dot{Q}_{wk} \rangle$ is the rate of fluid–solid interfacial heat transfer into phase k per unit volume of phase k. It is defined by

$$\gamma_v \alpha_k \, {}^{3i}\langle \dot{Q}_{wk} \rangle = -v^{-1} \int_{A_{wk}} \underline{J}_{qk} \cdot \underline{n}_k \, dA. \qquad (4.3.9)$$

Collectively, for most engineering problems, $\gamma_v \alpha_k ({}^{3i}\langle J_{Ek} \rangle + {}^{3i}\langle \dot{Q}_{kf} \rangle + {}^{3i}\langle \dot{Q}_{wk} \rangle)$ represents the distributed heat source and sink in v_k.

4.6.3.2 In terms of internal energy u_k

$$\gamma_v \frac{\partial}{\partial t} \alpha_k \, {}^{3i}\langle \rho_k u_k \rangle + \gamma_A \nabla \cdot \alpha_k \, {}^{2i}\langle \rho_k \underline{U}_k u_k \rangle$$
$$= -\gamma_v \alpha_k \, {}^{3i}\langle P_k \nabla \cdot \underline{U}_k \rangle - \gamma_A \nabla \cdot \alpha_k \, {}^{2i}\langle \underline{J}_{qk} \rangle$$
$$+ \gamma_v \alpha_k ({}^{3i}\langle \Phi_k \rangle + {}^{3i}\langle J_{Ek} \rangle + {}^{3i}\langle \dot{Q}_{kf} \rangle + {}^{3i}\langle \dot{Q}_{wk} \rangle)$$
$$- v^{-1} \int_{A_k} \rho_k u_k (\underline{U}_k - \underline{W}_k) \cdot \underline{n}_k \, dA, \qquad (4.6.4)$$

where Φ_k is the dissipation function given by

$$\Phi_k = \underline{\underline{\tau}}_k : \nabla, \underline{U}_k, \qquad (4.6.4a)$$

in which the double dot denotes the scalar product of two second-order tensors, and the comma denotes dyadic operation. Φ_k gives the dissipation rate per unit volume of phase k resulting from the irreversible conversion of mechanical work into thermal energy. Collectively, $\gamma_v \alpha_k ({}^{3i}\langle \Phi_k \rangle + {}^{3i}\langle J_{Ek} \rangle + {}^{3i}\langle \dot{Q}_{kf} \rangle + {}^{3i}\langle \dot{Q}_{wk} \rangle)$ represents the distributed heat source and sink in v_k.

4.6.3.3 In terms of enthalpy h_k

$$
\begin{aligned}
&\gamma_v \frac{\partial}{\partial t}\alpha_k {}^{3i}\langle \rho_k h_k \rangle + \gamma_A \nabla \cdot \alpha_k {}^{2i}\langle \rho_k \underline{U}_k h_k \rangle \\
&= \gamma_v \frac{\partial}{\partial t}\alpha_k {}^{3i}\langle P_k \rangle + \gamma_A \nabla \cdot \alpha_k {}^{2i}\langle \underline{U}_k P_k \rangle \\
&\quad - \gamma_v \alpha_k {}^{3i}\langle P_k \nabla \cdot \underline{U}_k \rangle - \gamma_A \nabla \cdot \alpha_k {}^{2i}\langle \underline{J}_{qk} \rangle \\
&\quad + \gamma_v \alpha_k \left({}^{3i}\langle \Phi_k \rangle + {}^{3i}\langle J_{Ek} \rangle + {}^{3i}\langle \dot{Q}_{kf} \rangle + {}^{3i}\langle \dot{Q}_{wk} \rangle \right) \\
&\quad + \upsilon^{-1} \int_{A_k} P_k \left(\underline{U}_k - \underline{W}_k \right) \cdot \underline{n}_k dA \\
&\quad - \upsilon^{-1} \int_{A_k} \rho_k h_k \left(\underline{U}_k - \underline{W}_k \right) \cdot \underline{n}_k dA
\end{aligned} \tag{4.6.5}
$$

Again, collectively $\gamma_v \alpha_k ({}^{3i}\langle \Phi_k \rangle + {}^{3i}\langle J_{Ek} \rangle + {}^{3i}\langle \dot{Q}_{kf} \rangle + {}^{3i}\langle \dot{Q}_{wk} \rangle)$ represents the distributed heat source and sink in υ_k.

4.7 Summary of local volume-averaged interfacial balance equations

In previous papers [2–5], the local volume-averaged interfacial balance equations were presented. For the convenience of the reader, they are listed as follows.

4.7.1 Local volume-averaged interfacial mass balance equation

$$
\begin{aligned}
&-\upsilon^{-1} \int_{A_k} \rho_k \left(\underline{U}_k - \underline{W}_k \right) \cdot \underline{n}_k dA = \gamma_v \alpha_k \Gamma_k \\
&= \upsilon^{-1} \int_{A_f} \rho_f (\underline{U}_f - \underline{W}_f) \cdot \underline{n}_f dA = -\gamma_v \alpha_f \Gamma_f,
\end{aligned} \tag{4.7.1}
$$

where the interfacial velocity $\underline{W}_k$ implies $\underline{W}_{kf}$ and the interfacial velocity $\underline{W}_f$ implies $\underline{W}_{fk}$. Because the interface has zero thickness, $\underline{W}_k = \underline{W}_f$ at any location of the interface. Clearly,

$$\gamma_v \alpha_k \Gamma_k + \gamma_v \alpha_f \Gamma_f = 0. \qquad (4.7.1a)$$

4.7.2 Local volume-averaged interfacial linear momentum balance equation

$$\begin{aligned} & v^{-1}\int_{A_k} (-P_k\underline{\underline{I}} + \underline{\underline{\tau}}_k)\cdot \underline{n}_k dA - v^{-1}\int_{A_k} \rho_k \underline{U}_k (\underline{U}_k - \underline{W}_k)\cdot \underline{n}_k dA \\ & = -v^{-1}\int_{A_f} (-P_f\underline{\underline{I}} + \underline{\underline{\tau}}_f)\cdot \underline{n}_f dA \\ & + v^{-1}\int_{A_f} \rho_f \underline{U}_f(\underline{U}_f - \underline{W}_f)\cdot \underline{n}_f dA \\ & - v^{-1}\int_{A_k} \left(-\nabla_{kf}\sigma_{kf} + 2\sigma_{kf}H_{kf}\underline{n}_k\right) dA \end{aligned} \qquad (4.7.2)$$

For bubbles and droplets, the last integral in Eq. (4.7.2) can be expressed in terms of capillary pressure difference:

$$v^{-1}\int_{A_k} \left(P_{ck} - P_{cf}\right)\underline{n}_k dA = v^{-1}\int_{A_k} \left(-\nabla_{kf}\sigma_{kf} + 2\sigma_{kf}H_{kf}\underline{n}_k\right) dA. \qquad (4.7.2a)$$

An equivalent expression can be written in terms of A_f, recognizing that for the interface between phases k and f, $A_k = A_f$, $\underline{n}_k = -\underline{n}_f$, and $H_{kf} = -H_{fk}$.

4.7.3 Local volume-averaged interfacial energy balance equation

The local volume-averaged interfacial energy balance equations are listed in this section.

4.7.3.1 Total energy balance (capillary energy ignored)

$$
\begin{aligned}
& v^{-1}\int_{A_k}\left(-P_k\underline{U}_k+\underline{\underline{\tau}}_k\cdot\underline{U}_k\right)\cdot\underline{n}_k dA \\
& \quad -v^{-1}\int_{A_k}\rho_k E_k\left(\underline{U}_k-\underline{W}_k\right)\cdot\underline{n}_k dA-v^{-1}\int_{A_k}\underline{J}_{qk}\cdot\underline{n}_k dA \\
& = -v^{-1}\int_{A_f}\left(-P_f\underline{U}_f+\underline{\underline{\tau}}_f\cdot\underline{U}_f\right)\cdot\underline{n}_f dA \\
& \quad +v^{-1}\int_{A_f}\rho_f E_f(\underline{U}_f-\underline{W}_f)\cdot\underline{n}_f dA+v^{-1}\int_{A_f}\underline{J}_{qf}\cdot\underline{n}_f dA
\end{aligned} \tag{4.7.3}
$$

4.7.3.2 Internal energy balance (dissipation and reversible work ignored)

$$
\begin{aligned}
& -v^{-1}\int_{A_k}\rho_k u_k\left(\underline{U}_k-\underline{W}_k\right)\cdot\underline{n}_k dA-v^{-1}\int_{A_k}\underline{J}_{qk}\cdot\underline{n}_k dA \\
& \quad = v^{-1}\int_{A_f}\rho_f u_f(\underline{U}_f-\underline{W}_f)\cdot\underline{n}_f dA+v^{-1}\int_{A_f}\underline{J}_{qf}\cdot\underline{n}_f dA
\end{aligned} \tag{4.7.4}
$$

4.7.3.3 Enthalpy balance (capillary energy ignored)

$$
\begin{aligned}
& v^{-1}\int_{A_k}P_k\left(\underline{U}_k-\underline{W}_k\right)\cdot\underline{n}_k dA \\
& \quad -v^{-1}\int_{A_k}\rho_k h_k\left(\underline{U}_k-\underline{W}_k\right)\cdot\underline{n}_k dA-v^{-1}\int_{A_k}\underline{J}_{qk}\cdot\underline{n}_k dA \\
& = -v^{-1}\int_{A_f}P_f(\underline{U}_f-\underline{W}_f)\cdot\underline{n}_f dA \\
& \quad +v^{-1}\int_{A_f}\rho_f h_f(\underline{U}_f-\underline{W}_f)\cdot\underline{n}_f dA+v^{-1}\int_{A_f}\underline{J}_{qf}\cdot\underline{n}_f dA
\end{aligned} \tag{4.7.5}
$$

5 Time averaging of local volume-averaged conservation equations or time-volume-averaged conservation equations and interfacial balance equations

The local volume-averaged multiphase conservation equations given in Chapter 4 are differential-integral equations. Before they can be used for either further analysis or numerical computation, it is necessary to (1) to express the volume averages of the product of the dependent variables in terms of the product of their volume averages and (2) to evaluate the interfacial transfer integrals that depend on the local values of the dependent variables at every point on the interface. To this end, we postulated that a point-dependent variable ψ_k for phase k can be expressed as the sum of its local intrinsic volume average ${}^{3i}\langle\psi_k\rangle$, and a spatial deviation $\tilde{\psi}_k$. ψ_k can be a scalar, vector, or tensor.

5.1 Basic postulates

It is postulated [6] that both ${}^{3i}\langle\psi_k\rangle$ and $\tilde{\psi}_k$ have a low-frequency component to be denoted by the subscript LF

and a high-frequency component to be denoted by a prime. Thus,

$$\psi_k = {}^{3i}\langle\psi_k\rangle + \tilde{\psi}_k \tag{5.1.1a}$$

$$\begin{aligned}\psi_k &= {}^{3i}\langle\psi_k\rangle_{LF} + {}^{3i}\langle\psi_k\rangle' + \tilde{\psi}_{kLF} + \tilde{\psi}'_k \\ &= {}^{3i}\langle\psi_k\rangle_{LF} + \tilde{\psi}_{kLF} + {}^{c}\psi'_k,\end{aligned} \tag{5.1.1b}$$

where

$${}^{c}\psi'_k = {}^{3i}\langle\psi_k\rangle' + \tilde{\psi}'_k. \tag{5.1.2}$$

The superscript c is a reminder that ${}^{c}\psi'_k$ is a composite of two high-frequency fluctuations. The low-frequency component refers to one that is a slowly varying function of time, including the time-dependent limiting case. The high-frequency component varies rapidly with time.

The time that characterizes the low-frequency component is of the order of

$$\tau_{LF} = \frac{L_c}{(\Delta U)_c} = \frac{\text{characteristic dimension of physical system}}{\text{characteristic low-frequency speed variation at a typical location}}. \tag{5.1.3a}$$

The characteristic time of the high-frequency component is of the order of

$$\begin{aligned}\tau_{HF} &= \frac{\Lambda}{(rms\,U')} \\ &= \frac{\text{characteristic length scale of high-frequency fluctuation}}{\text{root mean square of fluctuating velocity or turbulence intensity}} \\ &= \frac{1}{\text{characteristic spectral frequncy}}.\end{aligned} \tag{5.1.3b}$$

When time averaging is performed, the duration T over which the averaging is to be made must satisfy the following inequality:

$$\tau_{HF} \ll T \ll \tau_{LF}. \tag{5.1.4}$$

The spatial decomposition of the form given by Eq. (5.1.1a) was first suggested by Gray [21]. When the length-scale inequalities [Eq. (2.4.3)] are satisfied, the length scales associated with ${}^{3i}\langle\psi_k\rangle$ and $\tilde{\psi}_k$ are separable. The same is true for ${}^{3i}\langle\psi_k\rangle_{LF}$ and $\tilde{\psi}_{kLF}$. When Eq. (5.1.4) for the time-scale inequalities is satisfied, quantities with subscript LF and those denoted by a prime are also separable in the time or frequency domain. When the two characteristic times τ_{LF} and τ_{HF} overlap, such separation will not be possible. However, in practical applications, distinctions are usually feasible. Examples are: Duct flow with turbulence, a bubbly liquid in turbulent motion where the bubble phase configuration responds to low-frequency pressure fluctuation, and the case of impulsive motion produced by sudden break (LOCA) where high-frequency wave motion might not be important [34]. In the following derivations, high- and low-frequency parts are assumed to be separable.

If one adopts the Reynolds hypothesis used in elementary turbulence analysis, then the point instantaneous variables ψ_k can be decomposed as

$$\psi_k = {}^{t}\langle\psi_k\rangle + \psi_k' = \psi_{kLF} + \psi_k'. \tag{5.1.5}$$

Here, ${}^{t}\langle\psi_k\rangle$ denotes the temporal mean or a low-frequency component ψ_{kLF} and ψ_k' denotes the high-frequency fluctuating component. The time average ${}^{t}\langle\psi_k\rangle$ is defined by

$$ {}^{t}\langle\psi_k\rangle = \frac{1}{T}\int_{-T/2}^{T/2}\psi_k dt. \tag{5.1.6}$$

Comparing Eq. (5.1.1b) with Eq. (5.1.5) leads to the conclusion that

$$ {}^{c}\psi'_k = \psi'_k, \tag{5.1.7}$$

as one would intuitively expect. Hence, although ${}^{3i}\langle\psi_k\rangle'$ and $\tilde{\psi}'_k$ are not local entities, their sum is a point quantity. Substituting Eq. (5.1.7) into Eq. (5.1.1b) gives

$$\psi_k = {}^{3i}\langle\psi_k\rangle_{LF} + \tilde{\psi}_{kLF} + \psi'_k. \tag{5.1.8}$$

It is used again to demonstrate that ${}^{3i}\langle\psi_k\rangle_{LF}$ closely approximates ${}^{3i}\langle\psi_{kLF}\rangle$ as shown in Eq. (5.1.11).

Taking the intrinsic local volume averages of Eqs. (5.1.5) and (5.1.8), one obtains

$$ {}^{3i}\langle\psi_k\rangle = {}^{3i}\langle\psi_{kLF}\rangle + {}^{3i}\langle\psi'_k\rangle \tag{5.1.9a}$$

and

$$ {}^{3i}\langle\psi_k\rangle = {}^{3i}\langle\psi_k\rangle_{LF} + {}^{3i}\langle\tilde{\psi}_{kLF}\rangle + {}^{3i}\langle\psi'_k\rangle. \tag{5.1.9b}$$

Hence,

$$ {}^{3i}\langle\tilde{\psi}_{kLF}\rangle = 0 \tag{5.1.10}$$

$$ {}^{3i}\langle\psi_{kLF}\rangle \simeq {}^{3i}\langle\psi_k\rangle_{LF}. \tag{5.1.11}$$

Although scalars ρ_k, P_k, E_k, u_k, and h_k; vectors $\underline{U}_k$ and $\underline{J}_{qk}$; and tensor $\underline{\underline{\tau}}_k$ are to be decomposed in accordance with

Eq. (5.1.8), the local volume fraction α_k of phase k may be represented by

$$\alpha_k = \alpha_{kLF} + \alpha'_k \tag{5.1.12}$$

because α_k is inherently a volume-averaged quantity. Except for extreme conditions, the magnitude of α'_k, υ'_k, and A'_k is often much smaller than the corresponding magnitude of point variables such as ρ'_k, P'_k, E'_k, u'_k, h'_k, $\underline{U}'_k$, $\underline{J}'_{qk}$, and $\underline{\underline{\tau}}'_k$. In addition, as pointed out previously, the present analysis is best suited for dispersed systems for which interfacial tension would normally play a prominent role and abrupt changes in surface curvature are not expected to occur. The familiar smooth and gentle shapes of oscillating bubbles and droplets are examples [35]. The analysis [36] also shows that if α'_k is deleted, then A'_k should also be set to zero. For simplicity and lacking definitive experimental data, we assume in the present analysis that α'_k, A'_k, and υ'_k are all negligible. Further justifications for assuming α'_k, A'_k, and υ'_k are all negligible are presented in Appendix E.

The interfacial velocity $\underline{W}'_k$ appears only as a point variable in the interfacial transfer integrals of the governing differential-integral conservation equations. Thus, it needs only to be decomposed as

$$\underline{W}_k = \underline{W}_{kLF} + \underline{W}'_k. \tag{5.1.13}$$

If the flow is such that the characteristic length of the dispersed phase d is large compared to the integral length scale of turbulence Λ, and if there is no vigorous interfacial mass transfer, then the high-frequency fluctuating component of the interface velocity $\underline{W}'_k$ would not be significant and should be deleted. However, if $d \ll \Lambda$, then $\underline{W}'_k$ may not be ignored even though $A'_k = 0$. The case of turbulent flow of suspension of small solid particles in a gas or liquid is an example. Hence, $\underline{W}'_k$ is retained in the analysis that follows.

5.2 Useful observation without assuming $v'_k = 0$

The local volume average of a fluid property ψ_k of phase k is related to its intrinsic average according to

$$ {}^{3}\langle \psi_k \rangle = \gamma_v \alpha_k \, {}^{3i}\langle \psi_k \rangle, \qquad (5.2.1) $$

where γ_v is the volume porosity and α_k is the volume fraction of phase k in the fluid mixture. When the flow is turbulent, ψ_k may be decomposed into a low-frequency component ψ_{kLF} and a high-frequency component ψ'_k, as has been widely adopted in elementary turbulence analysis. Thus,

$$ \psi_k = \psi_{kLF} + \psi'_k. \qquad (5.2.2) $$

Similar expressions can be written for v_k, α_k, and so forth. We now compare the intrinsic volume average of the LF component of ψ_k, ${}^{3i}\langle \psi_{kLF} \rangle$, and the LF component of the intrinsic volume average of ψ_k, ${}^{3i}\langle \psi_k \rangle_{LF}$,

$$ {}^{3i}\langle \psi_{kLF} \rangle = \frac{1}{v \gamma_v \alpha_k} \int_{v_{kLF}} \psi_{kLF} dv \qquad (5.2.3a) $$

because $\int_{v'_k} \psi_{kLF} dv = 0$ as $\psi_{kLF} = 0$ in v'_k.

Denoting $\frac{v'_k}{v_{kLF}}$ by ε, we can write Eq. (5.2.3a) as

$$ {}^{3i}\langle \psi_{kLF} \rangle = \frac{1}{v_{kLF}(1+\varepsilon)} \int_{v_{kLF}} \psi_{kLF} dv. \qquad (5.2.3b) $$

Now,

$$
\begin{aligned}
{}^{3i}\langle \psi_k \rangle_{LF} &= \left[\frac{1}{\upsilon \gamma_{\upsilon} \alpha_k} \int_{\upsilon_{kLF}+\upsilon_k'} (\psi_{kLF} + \psi_k') d\upsilon \right]_{LF} \\
&= \frac{1}{\upsilon_{kLF}} \int_{\upsilon_{kLF}} \psi_{kLF} d\upsilon, \qquad (5.2.4)
\end{aligned}
$$

where the LF terms inside the bracket are sorted out. Dividing Eq. (5.2.4) by Eq. (5.2.3b) leads to

$$
\frac{{}^{3i}\langle \psi_k \rangle_{LF}}{{}^{3i}\langle \psi_{kLF} \rangle} = 1 + \varepsilon. \qquad (5.2.5)
$$

When $\varepsilon \ll 1$, the two intrinsic averages are approximately identical.

5.3 Time-volume-averaged mass conservation equation

The local volume-averaged mass conservation is given by Eq. (4.6.1):

$$
\begin{aligned}
&\gamma_{\upsilon} \frac{\partial}{\partial t} \alpha_k {}^{3i}\langle \rho_k \rangle + \gamma_{A} \nabla \cdot \alpha_k {}^{2i}\langle \rho_k \underline{U}_k \rangle \\
&= -\upsilon^{-1} \int_{A_k} \rho_k \left(\underline{U}_k - \underline{W}_k \right) \cdot \underline{n}_k dA. \qquad (4.6.1)
\end{aligned}
$$

Time averaging of the local volume-averaged mass conservation equation requires consideration of the following.

The first term on the left-hand side (LHS) of Eq. (4.6.1) is

$$
\gamma_{\upsilon} \frac{\partial}{\partial t} {}^{t}\langle \alpha_k {}^{3i}\langle \rho_k \rangle \rangle. \qquad (5.3.1)
$$

Because $\rho_k = {}^{3i}\langle\rho_k\rangle_{LF} + \tilde{\rho}_{kLF} + \rho_k'$, ${}^{3i}\langle\rho_k\rangle = {}^{3i}\langle\rho_k\rangle_{LF} + {}^{3i}\langle\rho_k'\rangle$,

$$\alpha_k\,{}^{3i}\langle\rho_k\rangle = \alpha_k\,{}^{3i}\langle\rho_k\rangle_{LF} + \alpha_k\,{}^{3i}\langle\rho_k'\rangle \qquad (5.3.1a)$$

$$^{t}\langle\alpha_k\,{}^{3i}\langle\rho_k\rangle\rangle = \alpha_k\,{}^{3i}\langle\rho_k\rangle_{LF}. \qquad (5.3.1b)$$

The first term on the LHS of Eq. (4.6.1) becomes

$$\gamma_v \frac{\partial}{\partial t}{}^{t}\langle\alpha_k\,{}^{3i}\langle\rho_k\rangle\rangle = \gamma_v \frac{\partial \alpha_k\,{}^{3i}\langle\rho_k\rangle_{LF}}{\partial t}. \qquad (5.3.1c)$$

The second term on the LHS of Eq. (4.6.1) is

$$\gamma_A \nabla : {}^{t}\langle\alpha_k\,{}^{2i}\langle\rho_k \underline{U}_k\rangle\rangle. \qquad (5.3.2)$$

$$\text{Because } \rho_k = {}^{2i}\langle\rho_k\rangle_{LF} + \tilde{\rho}_{kLF} + \rho_k',$$
$$\underline{U}_k = {}^{2i}\langle\underline{\tilde{U}}_k\rangle_{LF} + \underline{\tilde{U}}_{kLF} + \underline{U}_k' \qquad (5.3.2a)$$

$$\begin{aligned}{}^{2i}\langle\rho_k\underline{U}_k\rangle = {}& {}^{2i}\langle\rho_k\rangle_{LF}\,{}^{2i}\langle\underline{U}_k\rangle_{LF} + {}^{2i}\langle\rho_k\rangle_{LF}\,{}^{2i}\langle\underline{\tilde{U}}_{kLF}\rangle \\ &+ {}^{2i}\langle\rho_k\rangle_{LF}\,{}^{2i}\langle\underline{U}_k'\rangle + {}^{2i}\langle\tilde{\rho}_{kLF}\rangle\,{}^{2i}\langle\underline{U}_k\rangle_{LF} \\ &+ {}^{2i}\langle\tilde{\rho}_{kLF}\underline{\tilde{U}}_{kLF}\rangle + {}^{2i}\langle\tilde{\rho}_{kLF}\underline{U}_k'\rangle + {}^{2i}\langle\rho_k'\rangle\,{}^{2i}\langle\underline{U}_k\rangle_{LF} \\ &+ {}^{2i}\langle\rho_k'\underline{\tilde{U}}_{kLF}\rangle + {}^{2i}\langle\rho_k'\underline{U}_k'\rangle. \end{aligned} \qquad (5.3.2b)$$

$$\text{Because } {}^{2i}\langle{}^{2i}\langle\rho_k\rangle_{LF}\underline{\tilde{U}}_{kLF}\rangle = {}^{2i}\langle\rho_k\rangle_{LF}\,{}^{2i}\langle\underline{\tilde{U}}_k\rangle_{LF} = 0,$$
$${}^{2i}\langle\tilde{\rho}_{kLF}\,{}^{2i}\langle\underline{U}_k\rangle_{LF}\rangle = {}^{2i}\langle\tilde{\rho}_{kLF}\rangle\,{}^{2i}\langle\underline{U}_k\rangle_{LF} = 0 \qquad (5.3.2c)$$

$$\begin{aligned}{}^{2i}\langle\rho_k\underline{U}_k\rangle = {}& {}^{2i}\langle\rho_k\rangle_{LF}\,{}^{2i}\langle\underline{U}_k\rangle_{LF} + {}^{2i}\langle\rho_k\rangle_{LF}\,{}^{2i}\langle\underline{U}_k'\rangle \\ &+ {}^{2i}\langle\tilde{\rho}_{kLF}\underline{\tilde{U}}_{kLF}\rangle + {}^{2i}\langle\tilde{\rho}_{kLF}\underline{U}_k'\rangle \\ &+ {}^{2i}\langle\rho_k'\rangle\,{}^{2i}\langle\underline{U}_k\rangle_{LF} + {}^{2i}\langle\rho_k'\underline{\tilde{U}}_{kLF}\rangle + {}^{2i}\langle\rho_k'\underline{U}_k'\rangle. \end{aligned} \qquad (5.3.2d)$$

Hence,

$$\begin{aligned}{}^{t}\langle\alpha_k\,{}^{2i}\langle\rho_k\underline{U}_k\rangle\rangle = {}& \alpha_k({}^{2i}\langle\rho_k\rangle_{LF}\,{}^{2i}\langle\underline{U}_k\rangle_{LF} \\ &+ {}^{2i}\langle\tilde{\rho}_{kLF}\underline{\tilde{U}}_{kLF}\rangle + {}^{t2i}\langle\rho_k'\underline{U}_k'\rangle). \end{aligned} \qquad (5.3.2e)$$

By defining the volume-averaged eddy diffusivity for mass transfer, D_{mk}^T, according to

$$\alpha_k^{t2i}\langle \rho_k' \underline{U}_k' \rangle = -D_{mk}^T \nabla \alpha_k^{2i} \langle \rho_k \rangle_{LF}, \tag{5.3.2f}$$

where ρ_k = constant, $\tilde{\rho}_{kLF} = \rho_k' = 0$, and D_{mk}^T vanishes.

By introducing Eq. (5.3.2f) into Eq. (5.3.2e), the second term on the LHS of Eq. (4.6.1) can be written as follows:

$$\begin{aligned} &\gamma_A \nabla \cdot {}^t\langle \alpha_k^{2i} \langle \rho_k \underline{U}_k \rangle \rangle \\ &\quad = \gamma_A \nabla \cdot (\alpha_k^{2i} \langle \rho_k \rangle_{LF}^{2i} \langle \underline{U}_k \rangle_{LF} + \underline{\psi}_{mk}^{2i}), \end{aligned} \tag{5.3.2g}$$

in which $\underline{\psi}_{mk}^{2i}$ is a mass flux vector defined by

$$\underline{\psi}_{mk}^{2i} = \alpha_k^{2i} \langle \tilde{\rho}_{kLF} \underline{\tilde{U}}_{kLF} \rangle + \alpha_k^{t2i} \langle \rho_k' \underline{U}_k' \rangle \tag{5.3.2h}$$

substituting Eq. (5.3.2f) into Eq. (5.3.2h)

$$\underline{\psi}_{mk}^{2i} = \alpha_k^{2i} \langle \tilde{\rho}_{kLF} \underline{\tilde{U}}_{kLF} \rangle - D_{mk}^T \nabla \alpha_k^{2i} \langle \rho_k \rangle_{LF}. \tag{5.3.2i}$$

The physical meaning of $\gamma_A \nabla \cdot^t \langle \alpha_k^{2i} \langle \rho_k \underline{U}_k \rangle \rangle$ and so forth in a staggered-grid computational system is given in Appendix A. When variables such as ρ_k, $\underline{U}_k$, $\rho_k \underline{U}_k$, and so forth are preceded with γ_A, the value of these variables is evaluated at the surface of a computational cell under consideration. Because ρ_k is calculated at the center of the computational cell, its value can be computed by using an "appropriated" averaging of two neighboring cells of the surface or using the upwind cell density knowing the velocity at the computational cell surface. Likewise, when variables such as ρ_k, $\underline{U}_k$, $\rho_k \underline{U}_k$, and so forth are preceded with γ_v, the value of these variables are evaluated at the center of a computational cell under consideration. Because ρ_k is calculated at the center of the computational cell, but $\underline{U}_k$ is not, its value can be computed by using an "appropriated" averaging of two neighboring surface velocities of the computational cell. Noted that both γ_{υ} and γ_{A} are invariant in time and space.

We now evaluate the time average of the total interfacial mass generation integral for phase k within υ.

The first term on the right hand side (RHS) of Eq. (4.6.1) can be written as follows:

$$
{}^{t}\left\langle -\upsilon^{-1}\int_{A_k}\rho_k\left(\underline{U}_k-\underline{W}_k\right)\cdot\underline{n}_k dA\right\rangle
= -\upsilon^{-1}\int_{A_k}{}^{t}\langle\rho_k\left(\underline{U}_k-\underline{W}_k\right)\rangle\cdot\underline{n}_k dA. \qquad (5.3.3)
$$

Because low- and high-frequency parts of Eq. (5.3.3) can be separated, it is straightforward to demonstrate that

$$
\begin{aligned}
&{}^{t}\langle\rho_k(\underline{U}_k-\underline{W}_k)\rangle \\
&= {}^{t}\langle({}^{3i}\langle\rho_k\rangle_{LF}+\tilde{\rho}_{kLF})({}^{3i}\langle\underline{U}_k\rangle_{LF}+\underline{\tilde{U}}_{kLF}-\underline{W}_{kLF})\rangle \\
&\quad+{}^{t}\langle\rho_k'(\underline{U}_k'-\underline{W}_k')\rangle+{}^{t}\langle({}^{3i}\langle\rho_k\rangle_{LF}+\tilde{\rho}_{kLF})(\underline{U}_k'-\underline{W}_k')\rangle \\
&\quad+{}^{t}\langle\rho_k'({}^{3i}\langle\underline{U}_k\rangle_{LF}+\underline{\tilde{U}}_{kLF}-\underline{W}_{kLF})\rangle. \qquad (5.3.3a)
\end{aligned}
$$

Consequently, Eq. (5.3.3a) can be written after time averaging as follows:

$$
\begin{aligned}
&-\upsilon^{-1}\int_{A_k}{}^{t}\langle\rho_k\left(\underline{U}_k-\underline{W}_k\right)\rangle\cdot\underline{n}_k dA \\
&=\gamma_{\upsilon}\,{}^{3i}\langle\rho_k\rangle_{LF}\left(\frac{\partial\alpha_k}{\partial t}+{}^{3i}\langle\underline{U}_k\rangle_{LF}\cdot\nabla\alpha_k\right) \\
&\quad-\upsilon^{-1}\,{}^{3i}\langle\rho_k\rangle_{LF}\int_{A_k}\underline{\tilde{U}}_{kLF}\cdot\underline{n}_k dA
\end{aligned}
$$

$$- \upsilon^{-1} \int_{A_k} \tilde{\rho}_{kLF} ({}^{3i}\langle \underline{U}_k \rangle_{LF} + \underline{\tilde{U}}_{kLF} - \underline{W}_{kLF}) \cdot \underline{n}_k dA$$

$$- \upsilon^{-1} {}^{t}\left\langle \int_{A_k} \rho'_k (\underline{U}'_k - \underline{W}'_k) \cdot \underline{n}_k dA \right\rangle. \qquad (5.3.3b)$$

In deriving Eq. (5.3.3b), we employed Eqs. (2.4.8) and (2.4.9).

Thus, the first term on the RHS of Eq. (4.6.1), the time average of the interfacial mass generation rate of phase k per unit volume in υ, is

$$\gamma_\upsilon \alpha_k {}^{t}\langle \Gamma_k \rangle = -\upsilon^{-1} \int_{A_k} {}^{t}\langle \rho_k (\underline{U}_k - \underline{W}_k) \rangle \cdot \underline{n}_k dA$$

$$= \gamma_\upsilon {}^{3i}\langle \rho_k \rangle_{LF} \left(\frac{\partial \alpha_k}{\partial t} + {}^{3i}\langle \underline{U}_k \rangle_{LF} \cdot \nabla \alpha_k \right) + (MTI)_k, \qquad (5.3.3c)$$

in which $(MTI)_k$ stands for the interfacial mass transfer integral defined by

$$(MTI)_k = -\upsilon^{-1}\, {}^{3i}\langle \rho_k \rangle_{LF} \int_{A_k} \underline{\tilde{U}}_{kLF} \cdot \underline{n}_k dA$$

$$- \upsilon^{-1} \int_{A_k} \tilde{\rho}_{kLF} ({}^{3i}\langle \underline{U}_k \rangle_{LF} + \underline{\tilde{U}}_{kLF} - \underline{W}_{kLF}) \cdot \underline{n}_k dA$$

$$- \upsilon^{-1} \int_{A_k} {}^{t}\langle \rho'_k (\underline{U}'_k - \underline{W}'_k) \rangle \cdot \underline{n}_k dA. \qquad (5.3.3d)$$

More work needs to be done to examine the relative importance of the various terms in Eq. (5.3.3c). It is further noted that the first term on the RHS of Eq. (5.3.3c) can be written as $\gamma_\upsilon {}^{3i}\langle \rho_k \rangle_{LF} \frac{d\alpha_k}{dt_k}$, where the substantive time derivative $\frac{d}{dt_k}$ is defined by

$$\frac{d}{dt_k} = \left(\frac{\partial}{\partial t} + {}^{3i}\langle \underline{U}_k \rangle_{LF} \cdot \nabla \right). \qquad (5.3.3e)$$

Performing time averaging of Eq. (4.6.1), followed by introducing the results given in Eqs. (5.3.1c), (5.3.2g), (5.3.3c), and (5.3.3d), leads to the desired time-volume-averaged mass conservation equation:

$$\gamma_v \frac{\partial}{\partial t}\alpha_k {}^{3i}\langle \rho_k \rangle_{LF} + \gamma_A \nabla \cdot \alpha_k {}^{2i}\langle \rho_k \rangle_{LF} {}^{2i}\langle \underline{U}_k \rangle_{LF} + \gamma_A \nabla \cdot \underline{\psi}^{2i}_{mk} = \gamma_v \alpha_k {}^{t}\langle \Gamma_k \rangle. \quad (5.3.4)$$

When $\rho_k =$ constant,

$$^{3i}\langle \underline{\rho}_k \rangle_{LF} = \rho_k, \tilde{\rho}_{kLF} = \rho'_k = 0 \quad (5.3.5)$$

$$^{o}\langle \underline{\psi}^{2i}_{mk} \rangle = 0. \quad (5.3.5a)$$

In this case, the time-volume-averaged mass conservation equation simplifies to

$$\gamma_A \nabla \cdot \alpha_k {}^{2i}\langle \underline{U}_k \rangle_{LF} = \gamma_v {}^{3i}\langle \underline{U}_k \rangle_{LF} \cdot \nabla \alpha_k + {}^{o}\langle MTI \rangle_k \quad (5.3.5b)$$

$$^{o}\langle MTI \rangle_k = -\upsilon^{-1} \int_{A_k} \underline{\tilde{U}}_{kLF} \cdot \underline{n}_k dA. \quad (5.3.5c)$$

It is interesting to note that for a single phase with constant density without internal stationary structure, $\alpha_k = \gamma_v = \gamma_A = 1.0$ and $A_k = 0.0$, then Eqs. (5.3.5b) and (5.3.5c) reduce to the following equation as expected:

$$\nabla \cdot \underline{U}_k = 0. \quad (5.3.5d)$$

5.4 Time-volume-averaged interfacial mass balance equation

The local volume-averaged mass balance equation for interface A_{kf} is given by Eq. (4.7.1). Using Eq. (5.3.3c), we have

$$\gamma_v {}^{3i}\langle \rho_k \rangle_{LF} \left(\frac{\partial \alpha_k}{\partial t} + {}^{3i}\langle \underline{U}_k \rangle_{LF} \cdot \nabla \alpha_k \right) + (MTI)_k + \gamma_v {}^{3i}\langle \rho_f \rangle_{LF} \left(\frac{\partial \alpha_f}{\partial t} + {}^{3i}\langle \underline{U}_f \rangle_{LF} \cdot \nabla \alpha_f \right) + (MTI)_f = 0. \quad (5.4.1)$$

Here, $(MTI)_k$ is defined in Eq. (5.3.3d):

$$(MTI)_f = -\upsilon^{-1\,3i}\langle\rho_f\rangle_{LF}\int_{A_f}\tilde{\underline{U}}_{fLF}\cdot\underline{n}_f dA$$
$$-\upsilon^{-1}\int_{A_f}\tilde{\rho}_{fLF}({}^{3i}\langle\underline{U}_f\rangle_{LF}+\tilde{\underline{U}}_{fLF}-\underline{W}_{fLF})\cdot\underline{n}_f dA$$
$$-\upsilon^{-1}\int_{A_f}{}^{t}\langle\rho'_f(\underline{U}'_f-\underline{W}'_f)\rangle\cdot\underline{n}_f dA. \qquad (5.4.2)$$

Following similar procedures in deriving the time-volume-averaged mass conservation equation and time-volume-averaged interfacial mass balance equation as presented in Sections 5.3 and 5.4, respectively, the time-volume-averaged linear momentum, total energy, internal energy, and enthalpy conservation equation, and time-volume-averaged interfacial linear momentum, total energy, internal energy, and enthalpy balance equation are presented in Sections 5.5 to 5.12.

5.5 Time-volume-averaged linear momentum conservation equation

The local volume-averaged linear momentum conservation equation is given by Eq. (4.6.2):

$$\gamma_\upsilon\frac{\partial}{\partial t}\alpha_k{}^{3i}\langle\rho_k\underline{U}_k\rangle+\gamma_A\nabla\cdot\alpha_k{}^{2i}\langle\rho_k\underline{U}_k\underline{U}_k\rangle$$
$$=-\gamma_\upsilon\nabla\alpha_k{}^{3i}\langle P_k\rangle+\gamma_A\nabla\cdot\alpha_k{}^{2i}\langle\underline{\underline{\tau}}_k\rangle+\gamma_\upsilon\alpha_k{}^{3i}\langle\rho_k\rangle\underline{f}$$
$$+\upsilon^{-1}\int_{A_k}(-P_k\underline{\underline{I}}+\underline{\underline{\tau}}_k)\cdot\underline{n}_k dA$$
$$-\upsilon^{-1}\int_{A_k}\rho\underline{U}_k(\underline{U}_k-\underline{W}_k)\cdot\underline{n}_k dA. \qquad (4.6.2)$$

Time averaging of the local volume-averaged momentum conservation equation requires consideration of the following.

The first term on the LHS of Eq. (4.6.2) is

$$\gamma_v \frac{\partial}{\partial t} {}^t\langle \alpha_k {}^{3i}\langle \rho \underline{U}_k \rangle\rangle \tag{5.5.1}$$

$$\begin{aligned} &{}^t\langle \alpha_k {}^{3i}\langle \rho_k \underline{U}_k \rangle\rangle \\ &= {}^t\langle {}^{3i}\langle \alpha_k ({}^{3i}\langle \rho_k \rangle_{LF} + \tilde{\rho}_{kLF} + \rho_k')({}^{3i}\langle \underline{U}_k \rangle_{LF} + \underline{\tilde{U}}_{kLF} + \underline{U}_k')\rangle\rangle \\ &= \alpha_k {}^{3i}\langle \rho_k \rangle_{LF} {}^{3i}\langle \underline{U}_k \rangle_{LF} + \underline{\psi}^{3i}_{mk}, \end{aligned} \tag{5.5.1a}$$

where $\underline{\psi}^{3i}_{mk}$ is mass flux vector that is similar to Eq. (5.3.2h) but evaluated at the center of computational cell for the problem under consideration:

$$\underline{\psi}^{3i}_{mk} = \alpha_k {}^{3i}\langle \tilde{\rho}_{kLF} \underline{\tilde{U}}_{kLF} \rangle + \alpha_k {}^{t3i}\langle \rho_k' \underline{U}_k' \rangle. \tag{5.5.1b}$$

The first term on the LHS can be written as follows:

$$\begin{aligned} &\gamma_v {}^t\left\langle \frac{\partial}{\partial t} \alpha_k {}^{3i}\langle \rho_k \underline{U}_k \rangle \right\rangle \\ &= \gamma_v \frac{\partial}{\partial t} \alpha_k {}^{3i}\langle \rho_k \rangle_{LF} {}^{3i}\langle \underline{U}_k \rangle_{LF} + \gamma_v \frac{\partial}{\partial t} \underline{\psi}^{3i}_{mk}. \end{aligned} \tag{5.5.1c}$$

The second term on the LHS of Eq. (4.6.2) is

$$\gamma_A \nabla \cdot {}^t\langle \alpha_k {}^{2i}\langle \rho_k \underline{U}_k \underline{U}_k \rangle\rangle \tag{5.5.2}$$

$$\begin{aligned} &{}^t\langle \alpha_k {}^{2i}\langle \rho_k \underline{U}_k \underline{U}_k \rangle\rangle \\ &= {}^t\langle \alpha_k {}^{2i}\langle ({}^{2i}\langle \rho_k \rangle_{LF} + \tilde{\rho}_{kLF} + \rho_k')({}^{2i}\langle \underline{U}_k \rangle_{LF} \\ &\quad + \underline{\tilde{U}}_{kLF} + \underline{U}_k')({}^{2i}\langle \underline{U}_k \rangle + \underline{\tilde{U}}_{kLF} + \underline{U}_k')\rangle\rangle \\ &= \alpha_k {}^{2i}\langle \rho_k \rangle_{LF} {}^{2i}\langle \underline{U}_k \rangle_{LF} {}^{2i}\langle \underline{U}_k \rangle_{LF} + 2\, {}^{2i}\langle \underline{U}_k \rangle_{LF} \underline{\psi}^{2i}_{mk} \\ &\quad - \alpha_k ({}^{2i}\langle \underline{\underline{\tau}}_k^T \rangle + {}^{2i}\langle \underline{\underline{\tilde{\tau}}}_k \rangle + {}^{2i}\langle \underline{\underline{\tilde{\tau}}}_k^T \rangle), \end{aligned} \tag{5.5.2a}$$

in which

(a) ${}^{2i}\langle \underline{\underline{\tau}}_k^T \rangle$ is the volume-averaged Reynolds stress tensor and defined as

$$ {}^{2i}\langle \rho_k \rangle_{LF} \, {}^{t2i}\langle \underline{U}'_k \underline{U}'_k \rangle + {}^{t2i}\langle \tilde{\rho}_{kLF} \underline{U}'_k \underline{U}'_k \rangle. \qquad (5.5.2b) $$

(b) ${}^{2i}\langle \underline{\underline{\tilde{\tau}}}_k \rangle$ is the volume-averaged dispersive stress tensor and defined as

$$ {}^{2i}\langle \rho_k \rangle_{LF} \, {}^{2i}\langle \underline{\tilde{U}}_{kLF} \underline{\tilde{U}}_{kLF} \rangle + {}^{2i}\langle \tilde{\rho}_{kLF} \underline{\tilde{U}}_{kLF} \underline{\tilde{U}}_{kLF} \rangle. \qquad (5.5.2c) $$

(c) ${}^{2i}\langle \underline{\underline{\tilde{\tau}}}_k^T \rangle$ is the volume-averaged turbulent dispersive stress tensor and defined as

$$ 2 \, {}^{t2i}\langle \underline{\tilde{U}}_{kLF} \rho'_k \underline{U}'_k \rangle. \qquad (5.5.2d) $$

(d) ${}^{t}\langle \rho'_k \underline{U}'_k \underline{U}'_k \rangle$ is a time correlation of the third order and assumed to be small; thus, it can be neglected. (5.5.2e)

The second term on the LHS of Eq. (4.6.2) can be written as follows:

$$ \begin{aligned} &\gamma_A \nabla \cdot {}^{t}\langle \alpha_k \, {}^{2i}\langle \rho_k \underline{U}_k \underline{U}_k \rangle \rangle \\ &= \gamma_A \nabla \cdot \big[\alpha_k \, {}^{2i}\langle \rho_k \rangle_{LF} \, {}^{2i}\langle \underline{U}_k \rangle_{LF} \, {}^{2i}\langle \underline{U}_k \rangle_{LF} \\ &\quad + 2 \, {}^{2i}\langle \underline{U}_k \rangle_{LF} \underline{\psi}^{2i}_{mk} \\ &\quad - \alpha_k \big({}^{2i}\langle \underline{\underline{\tau}}_k^T \rangle + {}^{2i}\langle \underline{\underline{\tilde{\tau}}}_k \rangle + {}^{2i}\langle \underline{\underline{\tilde{\tau}}}_k^T \rangle \big) \big]. \end{aligned} \qquad (5.5.2f) $$

The first term on the RHS of Eq. (4.6.2) is

$$-\gamma_v \nabla^t \langle \alpha_k{}^{3i} \langle P_k \rangle\rangle \tag{5.5.3}$$

$$\begin{aligned} {}^t\langle \alpha_k{}^{3i} \langle P_k \rangle\rangle &= {}^t\langle \alpha_k{}^{3i} ({}^{3i}\langle P_k \rangle_{LF} + \tilde{P}_{kLF} + P'_k)\rangle \\ &= \alpha_k{}^{3i} \langle P_k \rangle_{LF}. \end{aligned} \tag{5.5.3a}$$

The first term on the RHS of Eq. (4.6.2) can be written as follows:

$$-\gamma_v \nabla^t \langle \alpha_k{}^{3i} \langle P_k \rangle\rangle = -\gamma_v \nabla \alpha_k{}^{3i} \langle P_k \rangle_{LF}. \tag{5.5.3b}$$

The second term on the RHS of Eq. (4.6.2) is

$$\gamma_A \nabla \cdot {}^t\langle \alpha_k{}^{2i} \langle \underline{\underline{\tau}}_k \rangle\rangle \tag{5.5.4}$$

$$\begin{aligned} {}^t\langle \alpha_k{}^{2i} \langle \underline{\underline{\tau}}_k \rangle\rangle &= {}^t\langle \alpha_k{}^{2i} ({}^{2i}\langle \underline{\underline{\tau}}_k \rangle_{LF} + \underline{\underline{\tilde{\tau}}}_{kLF} + \underline{\underline{\tau}}'_k)\rangle \\ &= \alpha_k{}^{2i} \langle \underline{\underline{\tau}}_k \rangle_{LF}. \end{aligned} \tag{5.5.4a}$$

The second term on the RHS of Eq. (4.6.2) can be written as follows:

$$\gamma_A \nabla \cdot {}^t\langle \alpha_k{}^{2i} \langle \underline{\underline{\tau}}_k \rangle\rangle = \gamma_A \nabla \cdot \alpha_k{}^{2i} \langle \underline{\underline{\tau}}_k \rangle_{LF}. \tag{5.5.4b}$$

Here, $\underline{\underline{\tau}}_k$ is the viscous stress tensor for Newtonian fluids:

$$\begin{aligned} \underline{\underline{\tau}}_k = &\left(\lambda_k - \frac{2}{3}\mu_k\right) \nabla \cdot \underline{U}_k \underline{\underline{I}} + \mu_k [\nabla, \underline{U}_k \\ &+ (\nabla, \underline{U}_k)_c]. \end{aligned} \tag{5.5.4c}$$

Here, λ_k is the buck viscosity; μ_k is dynamic viscosity; $\nabla, \underline{U}_k$ is dyad; and subscript c denotes conjugate. Because λ_k and μ_k are independent of velocity gradient, Eq. (5.5.4c) gives following substituting the relation, $\underline{U}_k = {}^{2i}\langle \underline{U}_k \rangle_{LF} + \underline{\tilde{U}}_{kLF} + \underline{U}'_k$.

$$\begin{aligned}\underline{\underline{\tau}}_k &= \left(\lambda_k - \frac{2}{3}\mu_k\right)(\nabla \cdot {}^{2i}\langle \underline{U}_k \rangle_{LF})\underline{\underline{I}} \\ &+ \mu_k[\nabla, {}^{2i}\langle \underline{U}_k \rangle_{LF} + (\nabla, {}^{2i}\langle \underline{U}_k \rangle_{LF})_c] \\ &+ \left(\lambda_k - \frac{2}{3}\mu_k\right)(\nabla \cdot \underline{\tilde{U}}_{kLF})\underline{\underline{I}} \\ &+ \mu_k[\nabla, \underline{\tilde{U}}_{kLF} + (\nabla, \underline{\tilde{U}}_{kLF})_c] \\ &+ \left(\lambda_k - \frac{2}{3}\mu_k\right)(\nabla \cdot \underline{U}'_k)\underline{\underline{I}} \\ &+ \mu_k[\nabla, \underline{U}'_k + (\nabla, \underline{U}'_k)_c]. \end{aligned} \tag{5.5.4d}$$

For convenience in subsequent discussion, we express $\underline{\underline{\tau}}_k$ in the form

$$\underline{\underline{\tau}}_k = {}^{2i}\langle \underline{\underline{\tau}}_k \rangle_{LF} + \underline{\underline{\tilde{\tau}}}_{kLF} + \underline{\underline{\tau}}'_k. \tag{5.5.4e}$$

${}^{t}\langle {}^{2i}\langle \underline{\underline{\tau}}_k \rangle\rangle = {}^{2i}\langle \underline{\underline{\tau}}_k \rangle_{LF}$. It follows from Eq. (5.5.4d) that

$$\begin{aligned}{}^{2i}\langle \underline{\underline{\tau}}_k \rangle_{LF} &= \left(\lambda_k - \frac{2}{3}\mu_k\right)(\nabla \cdot {}^{2i}\langle \underline{U}_k \rangle_{LF})\,\underline{\underline{I}} \\ &+ \mu_k[\nabla, {}^{2i}\langle \underline{U}_k \rangle_{LF} + (\nabla, {}^{2i}\langle \underline{U}_k \rangle_{LF})_c] \\ &+ \left(\lambda_k - \frac{2}{3}\mu_k\right) {}^{2i}\langle \nabla \cdot \underline{\tilde{U}}_{kLF} \rangle \underline{\underline{I}} \\ &+ \mu_k[{}^{2i}\langle \nabla, \underline{\tilde{U}}_{kLF} \rangle + {}^{2i}\langle \nabla, \underline{\tilde{U}}_{kLF} \rangle_c] \end{aligned} \tag{5.5.4f}$$

$$\begin{aligned}\underline{\underline{\tilde{\tau}}}_{kLF} &= \left(\lambda_k - \frac{2}{3}\mu_k\right)(\nabla \cdot \underline{\tilde{U}}_{kLF} - {}^{2i}\langle \nabla \cdot \underline{\tilde{U}}_{kLF} \rangle)\underline{\underline{I}} \\ &+ \mu_k[\nabla, \underline{\tilde{U}}_{kLF} - {}^{2i}\langle \nabla, \underline{\tilde{U}}_{kLF} \rangle \\ &+ (\nabla, \underline{\tilde{U}}_{kLF})_c - {}^{2i}\langle \nabla, \underline{\tilde{U}}_{kLF} \rangle_c] \end{aligned} \tag{5.5.4g}$$

$$\begin{aligned}\underline{\underline{\tau}}'_k &= \left(\lambda_k - \frac{2}{3}\mu_k\right)(\nabla \cdot {}^{2i}\underline{U}'_k)\,\underline{\underline{I}} \\ &+ \mu_k[\nabla, \underline{U}'_k + (\nabla, \underline{U}'_k)_c]. \end{aligned} \tag{5.5.4h}$$

Clearly, ${}^{t}\langle \underline{\underline{\tau}}'_k \rangle = 0$. Also, ${}^{2i}\langle \tilde{\underline{\underline{\tau}}}_{kLF} \rangle = 0$, as it must. Summation of Eqs. (5.5.4f)–(5.5.4h) gives Eq. (5.5.4d), as it should.

We noted that ${}^{2i}\langle (\nabla, \tilde{\underline{U}}_{kLF})_c \rangle = {}^{2i}\langle \nabla, \tilde{\underline{U}}_{kLF} \rangle_c$; in other words, the volume average of the conjugate is the same as the conjugate of the volume average. The volume-averaged quantities with superscript $2i$ are evaluated at the surface of the computational cell. We note that the viscous stress tensor defined in Eq. (5.5.4f) is not to be confused with the Reynolds stresses. In highly turbulent flow, the viscous stresses are usually insignificant relative to the turbulent stresses; thus, they can be ignored in the momentum equation.

When the viscosities in Eq. (5.5.4c) are dependent on strain rates, the resulting expression ${}^{t}\langle {}^{3i}\langle \underline{\underline{\tau}}_k \rangle \rangle$ is complicated. Details are given in Appendix C.

The third term on the RHS of Eq. (4.6.2) is

$$\gamma_v {}^{t}\langle \alpha_k {}^{3i}\langle \rho_k \rangle \underline{f} \rangle, \tag{5.5.5}$$

where $\underline{f}$ is the field force per unit mass and is taken to be constant.

The third term on the RHS of Eq. (4.6.2) can be written as

$$\begin{aligned} \gamma_v {}^{t}\langle \alpha_k {}^{3i}\langle \rho_k \rangle \underline{f} \rangle &= \gamma_v {}^{t}\langle \alpha_k {}^{3i}\langle ({}^{3i}\langle \rho_k \rangle_{LF} + \tilde{\rho}_{kLF} + \rho'_k) \underline{f} \rangle \rangle \\ &= \gamma_v \alpha_k {}^{3i}\langle \rho_k \rangle_{LF} \underline{f}. \end{aligned} \tag{5.5.5a}$$

The fourth term on the RHS of Eq. (4.6.2) is

$$ {}^{t}\left\langle v^{-1} \int_{A_k} (-P_k \underline{\underline{I}} + \underline{\underline{\tau}}_k) \cdot \underline{n}_k dA \right\rangle = v^{-1} \int_{A_k} {}^{t}\langle -P_k \underline{\underline{I}} + \underline{\underline{\tau}}_k \rangle \cdot \underline{n}_k dA \tag{5.5.6}$$

$$\upsilon^{-1}\int_{A_k} {}^{t}\langle -({}^{3i}\langle P_k\rangle_{LF} + \tilde{P}_{kLF} + P_k')\underline{\underline{I}}\rangle \cdot \underline{n}_k dA$$
$$= \gamma_\upsilon{}^{3i}\langle P_k\rangle_{LF}\nabla\alpha_k - (PTI)_k \quad (5.5.6a)$$

$$\upsilon^{-1}\int_{A_k} {}^{t}\langle ({}^{3i}\langle \underline{\underline{\tau}}_k\rangle_{LF} + \tilde{\underline{\underline{\tau}}}_{kLF} + \underline{\underline{\tau}}_k') \cdot \underline{n}_k dA\rangle$$
$$= -\gamma_\upsilon{}^{3i}\langle \underline{\underline{\tau}}_k\rangle_{LF} \cdot \nabla\alpha_k + (VSTI)_k, \quad (5.5.6b)$$

where PTI and $VSTI$ denote the interfacial pressure transfer integral and the interfacial viscous stress transfer integral, respectively, and are defined as follows:

$$(PTI)_k = -\upsilon^{-1}\int_{A_k} \tilde{P}_{kLF}\underline{n}_k dA, \quad \text{and}$$
$$(VSTI)_k = -\upsilon^{-1}\int_{A_k} \tilde{\underline{\underline{\tau}}}_{kLF} \cdot \underline{n}_k dA. \quad (5.5.6c,d)$$

The fourth term can be written as sum of Eqs. (5.5.6a)–(5.5.6d).

$$\upsilon^{-1}\int_{A_k} {}^{t}\langle(-P_k\underline{\underline{I}} + \underline{\underline{\tau}}_k)\rangle \cdot \underline{n}_k dA$$
$$= \gamma_\upsilon{}^{3i}\langle P_k\rangle_{LF}\nabla\alpha_k - \gamma_\upsilon{}^{3i}\langle \underline{\underline{\tau}}_k\rangle_{LF} \cdot \nabla\alpha_k$$
$$+ (PTI)_k - (VSTI)_k. \quad (5.5.6e)$$

The fifth term on the RHS of Eq. (4.6.2) is the time-volume-averaged interfacial momentum transfer rate of phase k per unit volume:

$${}^{t}\left\langle -\upsilon^{-1}\int_{A_k} \rho_k \underline{U}_k(\underline{U}_k - \underline{W}_k) \cdot \underline{n}_k dA\right\rangle$$
$$= -\upsilon^{-1}\int_{A_k} {}^{t}\langle \rho_k \underline{U}_k(\underline{U}_k - \underline{W}_k)\rangle \cdot \underline{n}_k dA. \quad (5.5.7)$$

Because low- and high-frequency parts of Eq. (5.5.7) can be separated, it is straightforward to demonstrate that

$$
\begin{aligned}
{}^{t}\langle \rho_k \underline{U}_k (\underline{U}_k - \underline{W}_k) \rangle = {}^{t}\langle & ({}^{3i}\langle \rho_k \rangle_{LF} + \tilde{\rho}_{kLF})({}^{3i}\langle \underline{U}_k \rangle_{LF} + \underline{\tilde{U}}_{kLF}) \\
& \times ({}^{3i}\langle \underline{U}_k \rangle_{LF} + \underline{\tilde{U}}_{kLF} - \underline{W}_{kLF}) \\
& + \underline{U}'_k ({}^{3i}\langle \rho_k \rangle_{LF} + \tilde{\rho}_{kLF} + \rho'_k) \\
& \times ({}^{3i}\langle \underline{U}_k \rangle_{LF} + \underline{\tilde{U}}_{kLF} - \underline{W}_{kLF}) \\
& + \rho'_k ({}^{3i}\langle \underline{U}_k \rangle_{LF} + \underline{\tilde{U}}_{kLF}) \\
& \times ({}^{3i}\langle \underline{U}_k \rangle_{LF} + \underline{\tilde{U}}_{kLF} - \underline{W}_{kLF}) \\
& + ({}^{3i}\langle \rho_k \rangle_{LF} + \tilde{\rho}_{kLF}) \\
& \times ({}^{3i}\langle \underline{U}_k \rangle_{LF} + \underline{\tilde{U}}_{kLF} + \underline{U}'_k) (\underline{U}'_k - \underline{W}'_k) \\
& + \rho'_k ({}^{3i}\langle \underline{U}_k \rangle_{LF} + \underline{\tilde{U}}_{kLF})(\underline{U}'_k - \underline{W}'_k) \\
& + \rho'_k \underline{U}'_k (\underline{U}'_k - \underline{W}'_k) \rangle. \qquad (5.5.7a)
\end{aligned}
$$

Consequently, substituting Eq. (5.5.7a) into Eq. (5.5.7) and after the time averaging, we can write Eq. (5.5.7) as follows:

$$
\begin{aligned}
& -\upsilon^{-1} \int_{A_k} {}^{t}\langle \rho_k \underline{U}_k (\underline{U}_k - \underline{W}_k) \rangle \cdot \underline{n}_k dA \\
& = \gamma_\upsilon {}^{3i}\langle \rho_k \rangle_{LF} {}^{3i}\langle \underline{U}_k \rangle_{LF} \left(\frac{d\alpha_k}{dt} + {}^{3i}\langle \underline{U}_k \rangle_{LF} \cdot \nabla \alpha_k \right) \\
& \quad - \upsilon^{-1} {}^{3i}\langle \rho_k \rangle_{LF} {}^{3i}\langle \underline{U}_k \rangle_{LF} \int_{A_k} \underline{\tilde{U}}_{kLF} \cdot \underline{n}_k dA \\
& \quad - \upsilon^{-1} {}^{3i}\langle \underline{U}_k \rangle_{LF} \int_{A_k} \tilde{\rho}_{kLF} ({}^{3i}\langle \underline{U}_k \rangle_{LF} + \underline{\tilde{U}}_{kLF} - \underline{W}_{kLF}) \cdot \underline{n}_k dA \\
& \quad - \upsilon^{-1} \int_{A_k} ({}^{3i}\langle \rho_k \rangle_{LF} + \tilde{\rho}_{kLF}) \underline{\tilde{U}}_{kLF} \\
& \quad \times ({}^{3i}\langle \underline{U}_k \rangle_{LF} + \underline{\tilde{U}}_{kLF} - \underline{W}_{kLF}) \cdot \underline{n}_k dA
\end{aligned}
$$

$$- \upsilon^{-1} \int_{A_k} ({}^{3i}\langle \rho_k \rangle_{LF} + \tilde{\rho}_{kLF}) {}^{t}\langle \underline{U}'_k (\underline{U}'_k - \underline{W}'_k)\rangle \cdot \underline{n}_k dA$$
$$- \upsilon^{-1} \int_{A_k} ({}^{3i}\langle \underline{U}_k \rangle_{LF} + \underline{\tilde{U}}_{kLF}) {}^{t}\langle \rho'_k (\underline{U}'_k - \underline{W}'_k)\rangle \cdot \underline{n}_k dA$$
$$- \upsilon^{-1} {}^{t}\left\langle \int_{A_k} ({}^{3i}\langle \underline{U}_k \rangle_{LF} + \underline{\tilde{U}}_{kLF} - \underline{W}_{kLF}) {}^{t}\langle \rho'_k \underline{U}'_k \rangle \cdot \underline{n}_k dA \right\rangle$$
$$- \upsilon^{-1} \int_{A_k} {}^{t}\langle \rho'_k \underline{U}'_k (\underline{U}'_k - \underline{W}'_k)\rangle \cdot \underline{n}_k dA. \quad (5.5.7b)$$

The fifth term on the RHS of Eq. (4.6.2) can be written by multiplying Eq. (5.3.3c) by ${}^{3i}\langle \underline{U}_k \rangle_{LF}$, followed by introducing the result into Eq. (5.5.7b), we obtain the following:

$$-\upsilon^{-1} \int_{A_k} {}^{t}\langle \rho_k \underline{U}_k (\underline{U}_k - \underline{W}_k)\rangle \cdot \underline{n}_k dA$$
$$= \gamma_\upsilon \alpha_k {}^{t}\langle \Gamma_k \rangle {}^{3i}\langle \underline{U}_k \rangle_{LF} + (MMTI)_k \quad (5.5.7c)$$

The $(MMTI)_k$ stands for interfacial momentum transfer integral, which is defined by

$$(MMTI)_k = -\upsilon^{-1} \int_{A_k} [({}^{3i}\langle \rho_k \rangle_{LF} + \tilde{\rho}_{kLF}) \underline{\tilde{U}}_{kLF}]$$
$$\times ({}^{3i}\langle \underline{U}_k \rangle_{LF} + \underline{\tilde{U}}_{kLF} - \underline{W}_{kLF}) \cdot \underline{n}_k dA$$
$$- \upsilon^{-1} \int_{A_k} ({}^{3i}\langle \underline{U}_k \rangle_{LF} + \underline{\tilde{U}}_{kLF} - \underline{W}_{kLF}) {}^{t}\langle \rho'_k \underline{U}'_k \rangle$$
$$\cdot \underline{n}_k dA$$
$$- \upsilon^{-1} \int_{A_k} [({}^{3i}\langle \rho_k \rangle_{LF} + \tilde{\rho}_{kLF}) {}^{t}\langle \underline{U}'_k (\underline{U}'_k - \underline{W}'_k)\rangle$$
$$+ \underline{\tilde{U}}_{kLF} {}^{t}\langle \rho'_k (\underline{U}'_k - \underline{W}'_k)\rangle] \cdot \underline{n}_k dA$$
$$- \upsilon^{-1} \int_{A_k} {}^{t}\langle \rho'_k \underline{U}'_k (\underline{U}'_k - \underline{W}'_k)\rangle \cdot \underline{n}_k dA. \quad (5.5.7d)$$

One can see from Eqs. (5.5.6e), (5.5.4f), and (5.5.7c) that the total time-volume-averaged interfacial momentum transfer rate consists of two parts: (1) momentum transfer that results from the interfacial pressure and viscous stresses, and (2) momentum transfer that is directly related to interfacial mass generation and extraneous momentum transfer due to spatial deviation of velocity and density and due to time correlation of velocity and density. The time-averaged interfacial total momentum source of phase k per unit volume in υ, $\gamma_\upsilon \alpha_k{}^t\langle \underline{M}_k \rangle$, is

$$\begin{aligned}\gamma_\upsilon \alpha_k{}^t\langle \underline{M}_k \rangle = {} & \gamma_\upsilon \left({}^{3i}\langle P_k \rangle_{LF} \underline{\underline{I}} - {}^{3i}\langle \underline{\underline{\tau}}_k \rangle_{LF} \right) \cdot \nabla \alpha_k \\ & + (PTI)_k - (VSTI)_k \\ & + \gamma_\upsilon \alpha_k{}^t\langle \Gamma_k \rangle {}^{3i}\langle \underline{U}_k \rangle_{LF} + (MMTI)_k . \quad (5.5.7e)\end{aligned}$$

Using the results given in Eqs. (5.5.1c), (5.5.2f), (5.5.3b), (5.5.4b), (5.5.4f), (5.5.5a), (5.5.6e), and (5.5.7c), one obtains the time-volume-averaged linear momentum conservation equation:

$$\begin{aligned}& \gamma_\upsilon \frac{\partial}{\partial t} \alpha_k {}^{3i}\langle \rho_k \rangle_{LF} {}^{3i}\langle \underline{U}_k \rangle_{LF} + \gamma_\upsilon \frac{\partial}{\partial t} \underline{\psi}^{3i}_{mk} \\ & + \gamma_A \nabla \cdot \left[\alpha_k {}^{2i}\langle \rho_k \rangle_{LF} {}^{2i}\langle \underline{U}_k \rangle_{LF} {}^{2i}\langle \underline{U}_k \rangle_{LF} + 2\, {}^{2i}\langle \underline{U}_k \rangle_{LF} \underline{\psi}^{2i}_{mk} \right. \\ & \left. - \alpha_k \left({}^{2i}\langle \underline{\underline{\tau}}^T_k \rangle + {}^{2i}\langle \underline{\underline{\tilde{\tau}}}_k \rangle + {}^{2i}\langle \langle \underline{\underline{\tilde{\tau}}}^T_k \rangle \rangle \right) \right] = -\gamma_\upsilon \nabla \alpha_k {}^{3i}\langle P_k \rangle_{LF} \\ & + \gamma_A \nabla \cdot \alpha_k {}^{2i}\langle \underline{\underline{\tau}}_k \rangle_{LF} + \gamma_\upsilon \alpha_k {}^{3i}\langle \rho_k \rangle_{LF} \underline{f} \\ & + \gamma_\upsilon {}^{3i}\langle P_k \rangle_{LF} \nabla \alpha_k - \gamma_\upsilon {}^{3i}\langle \underline{\underline{\tau}}_k \rangle_{LF} \cdot \nabla \alpha_k + (PTI)_k - (VSTI)_k \\ & + \gamma_\upsilon \alpha_k{}^t\langle \Gamma_k \rangle {}^{3i}\langle \underline{U}_k \rangle_{LF} + (MMTI)_k . \qquad (5.5.7f)\end{aligned}$$

If we rearrange the terms on the LHS of Eq. (5.5.7f) and combine the first term and fourth term on the RHS of Eq. (5.5.7f), we get

$$\begin{aligned}
&\gamma_v \frac{\partial}{\partial t}\left(\alpha_k{}^{3i}\langle\rho_k\rangle_{LF}{}^{3i}\langle\underline{U}_k\rangle_{LF}+\underline{\psi}_{mk}^{3i}\right) \\
&\quad+\gamma_{A}\nabla\cdot\left[\alpha_k{}^{2i}\langle\rho_k\rangle_{LF}{}^{2i}\langle\underline{U}_k\rangle_{LF}{}^{2i}\langle\underline{U}_k\rangle_{LF}+2\,{}^{2i}\langle\underline{U}_k\rangle_{LF}\underline{\psi}_{mk}^{2i}\right. \\
&\quad\left.-\alpha_k\left({}^{2i}\langle\underline{\underline{\tau}}_k^T\rangle+{}^{2i}\langle\tilde{\underline{\underline{\tau}}}_k\rangle+{}^{2i}\langle\tilde{\underline{\underline{\tau}}}_k^T\rangle\right)\right] \\
&=-\gamma_v\alpha_k\nabla^{3i}\langle P_k\rangle_{LF}+\gamma_{A}\nabla\cdot\alpha_k{}^{2i}\langle\underline{\underline{\tau}}_k\rangle_{LF}+\gamma_v\alpha_k{}^{3i}\langle\rho_k\rangle_{LF}\underline{f} \\
&\quad-\gamma_v{}^{3i}\langle\underline{\underline{\tau}}_k\rangle_{LF}\cdot\nabla\alpha_k+(PTI)_k-(VSTI)_k \\
&\quad+\gamma_v\alpha_k{}^{t}\langle\Gamma_k\rangle^{3i}\langle\underline{U}_k\rangle_{LF}+(MMTI)_k\,.
\end{aligned} \tag{5.5.7g}$$

In some practical engineering applications, one may have experimentally measured the flow resistance or distributed resistance for the problem under consideration. In this case, it is advantageous to use the data, which could improve the accuracy of results and somewhat simplify the calculational procedure. For these reasons, we modify our formulation of the momentum equation to account for this situation:

1. The fourth term on the RHS of Eq. (4.6.2) can be split into two terms as follows:

$$\begin{aligned}
&\upsilon^{-1}\int_{A_k}\left(-P_k\underline{\underline{I}}+\underline{\underline{\tau}}_k\right)\cdot\underline{n}_k dA \\
&\quad=\upsilon^{-1}\int_{A_{kf}}\left(-P_k\underline{\underline{I}}+\underline{\underline{\tau}}_k\right)\cdot\underline{n}_k dA \\
&\qquad+\upsilon^{-1}\int_{A_{wk}}\left(-P_k\underline{\underline{I}}+\underline{\underline{\tau}}_k\right)\cdot\underline{n}_k dA
\end{aligned} \tag{5.5.7h}$$

and

$$\upsilon^{-1}\int_{A_{wk}}{}^{t}\langle-P_k\underline{\underline{I}}+\underline{\underline{\tau}}_k\rangle\cdot\underline{n}_k dA=-\gamma_v\alpha_k{}^{3i}\langle\underline{R}_k\rangle, \tag{5.5.7i}$$

where ${}^{3i}\langle \underline{R}_k \rangle$ is the distributed resistance force per unit volume of the phase k exerted by the stationary solid structures in υ. The distributed resistance force can be combined into the third term on the RHS of Eq. (4.6.2).

2. The integration limit A_k of Eqs. (5.5.6a)–(5.5.6d) must be replaced by A_{kf}.

With the preceding modifications, the time-volume-averaged linear momentum equation [Eq. (5.5.7f)] can be written as follows:

$$\begin{aligned}
&\gamma_\upsilon \frac{\partial}{\partial t}\left(\alpha_k {}^{3i}\langle \rho_k \rangle_{LF} {}^{3i}\langle \underline{U}_k \rangle_{LF} + \underline{\psi}^{3i}_{mk}\right) \\
&\quad + \gamma_A \nabla \cdot \left[\alpha_k {}^{2i}\langle \rho_k \rangle_{LF} {}^{2i}\langle \underline{U}_k \rangle_{LF} {}^{2i}\langle \underline{U}_k \rangle_{LF} \right. \\
&\quad \left. + 2\, {}^{2i}\langle \underline{U}_k \rangle_{LF} \underline{\psi}^{2i}_{mk} - \alpha_k \left({}^{2i}\langle \underline{\underline{\tau}}^T_k \rangle + {}^{2i}\langle \underline{\underline{\tilde{\tau}}}_k \rangle + {}^{2i}\langle \underline{\underline{\tilde{\tau}}}^T_k \rangle\right)\right] \\
&= -\gamma_\upsilon \nabla \alpha_k {}^{3i}\langle P_k \rangle_{LF} + \gamma_A \nabla \cdot \alpha_k {}^{2i}\langle \underline{\underline{\tau}}_k \rangle_{LF} \\
&\quad + \gamma_\upsilon \alpha_k \left({}^{3i}\langle \rho_k \rangle_{LF} \underline{f} - {}^{3i}\langle \underline{R}_k \rangle\right) \\
&\quad + \upsilon^{-1} \int_{A_{kf}} {}^{t}\langle -P_k \underline{\underline{I}} + \underline{\underline{\tau}}_k \rangle \cdot \underline{n}_k dA \\
&\quad + \gamma_\upsilon \alpha_k {}^{t}\langle \Gamma_k \rangle {}^{3i}\langle \underline{U}_k \rangle_{LF} + (MMTI)_k \,. \qquad (5.5.7\text{j})
\end{aligned}$$

The first integral on the RHS of Eq. (5.5.7h) can be evaluated as previously, except that the integral limit has changed from A_k to A_{kf}, and $\gamma_\upsilon \alpha_k {}^{3i}\langle \underline{R}_k \rangle$ is defined as Eq. (5.5.7i).

For Newtonian fluids, ${}^{2i}\langle \underline{\underline{\tau}}_k \rangle_{LF}$ is given by Eq. (5.5.4f), and the superscript $2i$ means that the shear stress tensor is evaluated at the surface of computational cell for the problem under consideration. The relative importance of various terms in Eqs. (5.5.7f), (5.5.7g), and (5.5.7j) remains to be assessed.

When $\rho_k =$ constant, ${}^{3i}\langle\rho_k\rangle_{LF} = \rho_k$, $\tilde{\rho}_{kLF} = \rho'_k = 0$. Therefore,

$$ {}^{o}\underline{\psi}^{3i}_{mk} = {}^{o}\underline{\psi}^{2i}_{mk} = 0 \tag{5.5.8} $$

$$ {}^{o2i}\langle\underline{\underline{\tau}}^{T}_{k}\rangle = \rho_k {}^{t}\langle\underline{U}'_k\underline{U}'_k\rangle \tag{5.5.8a} $$

$$ {}^{o2i}\langle\underline{\underline{\tilde{\tau}}}_{k}\rangle = \rho_k {}^{t}\langle\underline{\tilde{U}}_{kLF}\underline{\tilde{U}}_{kLF}\rangle \tag{5.5.8b} $$

$$ {}^{o2i}\langle\underline{\underline{\tilde{\tau}}}^{T}_{k}\rangle = 0 \tag{5.5.8c} $$

$$ \gamma_v\alpha_k {}^{ot}\langle\Gamma_k\rangle = \gamma_v\rho_k\left(\frac{\partial\alpha_k}{\partial t} + {}^{3i}\langle\underline{U}_k\rangle_{LF}\cdot\nabla\alpha_k\right) + {}^{o}(MTI)_k \tag{5.5.8d} $$

$$ {}^{o}(MTI)_k = -\upsilon^{-1}\rho_k\int_{A_k}\underline{\tilde{U}}_{kLF}\cdot\underline{n}_k dA \tag{5.5.8e} $$

$$ \begin{aligned} {}^{o}(MMTI)_k = -\upsilon^{-1}\rho_k\int_{A_k} & \underline{\tilde{U}}_{kLF}({}^{3i}\langle\underline{U}_k\rangle_{LF} \\ & + \underline{\tilde{U}}_{kLF} - \underline{W}_{kLF})\cdot\underline{n}_k dA \\ & - \upsilon^{-1}\rho_k\int_{A_k}{}^{t}\langle\underline{U}'_k(\underline{U}'_k - \underline{W}'_k)\rangle\cdot\underline{n}_k dA \end{aligned} \tag{5.5.8f} $$

$$ \gamma_v\alpha_k {}^{o3i}\langle\rho_k\rangle_{LF}\underline{f} = \gamma_v\alpha_k\rho_k\underline{f}. \tag{5.5.8g} $$

Substituting Eqs. (5.5.8)–(5.5.8g) into Eq. (5.5.7g), we can write the time-volume-averaged linear momentum equation for $\rho_k =$ constant as follows:

$$ \begin{aligned} & \gamma_v\rho_k\frac{\partial}{\partial t}\alpha_k {}^{3i}\langle\underline{U}_k\rangle_{LF} + \gamma_A\nabla\cdot\left[\alpha_k\rho_k {}^{2i}\langle\underline{U}_k\rangle_{LF} {}^{2i}\langle\underline{U}_k\rangle_{LF}\right. \\ & \quad \left. - \alpha_k\left({}^{o2i}\langle\underline{\underline{\tau}}^{T}_{k}\rangle + {}^{o2i}\langle\underline{\underline{\tilde{\tau}}}_{k}\rangle\right)\right] \\ & = -\gamma_v\alpha_k\nabla^{3i}\langle P_k\rangle_{LF} + \gamma_A\nabla\cdot\alpha_k {}^{2i}\langle\underline{\underline{\tau}}_k\rangle_{LF} + \gamma_v\alpha_k\rho_k\underline{f} \\ & \quad - \gamma_v {}^{3i}\langle\underline{\underline{\tau}}_k\rangle_{LF}\cdot\nabla\alpha_k + (PTI)_k - (VSTI)_k \\ & \quad + \gamma_v\alpha_k {}^{ot}\langle\Gamma_k\rangle {}^{3i}\langle\underline{U}_k\rangle_{LF} + {}^{o}(MMTI)_k. \end{aligned} \tag{5.5.9} $$

For any single-phase system without internal structure, $\alpha_k = 1$, $A_k = 0$, and $\gamma_v = \gamma_A = 1$. Therefore, all interfacial integrals vanished. If the system is at rest, then all quantities associated with $\underline{U}_k$ also vanished. Accordingly, Eqs. (5.5.7f), (5.5.7g), and (5.5.7j) reduce to

$$-\nabla^{3i}\langle P_k\rangle + {}^{3i}\langle \rho_k\rangle \underline{g} = 0, \tag{5.5.10}$$

with $\underline{f} = \underline{g}$, $\underline{g}$ being the gravitational acceleration vector. The subscript LF for ${}^{3i}\langle P_k\rangle$ and ${}^{3i}\langle \rho_k\rangle$ has been dropped, because the fluid is everywhere at rest. The characteristic length scale d in Eq. (2.4.3) is zero for a single-phase system; hence, the characteristic length scale of l can be made as small as desired. Thus, in the limit, ${}^{3i}\langle P_k\rangle \to P_k$ and ${}^{3i}\langle \rho_k\rangle \to \rho_k$, and Eqs. (5.5.7f), (5.5.7g), and (5.5.7j) become

$$-\nabla P_k + \rho_k \underline{g} = 0, \tag{5.5.10a}$$

thus satisfying the basic relation of fluid hydrostatics.

5.6 Time-volume-averaged interfacial linear momentum balance equation

Based on the local volume-averaged linear momentum balance equations [Eqs. (4.7.2) and (4.7.2a)], the time-volume-averaged interfacial linear momentum balance equation can readily be obtained by using Eq. (5.5.7e):

$$\begin{aligned}
&\gamma_v\left({}^{3i}\langle P_k\rangle_{LF}\underline{\underline{I}} - {}^{3i}\langle\underline{\underline{\tau}}_k\rangle_{LF}\right)\cdot\nabla\alpha_k + (PTI)_k \\
&\quad - (VSTI)_k + \gamma_v\alpha_k{}^t\langle\Gamma_k\rangle^{3i}\langle\underline{U}_k\rangle_{LF} + (MMTI)_k \\
&\quad + \gamma_v\left({}^{3i}\langle P_f\rangle_{LF}\underline{\underline{I}} - {}^{3i}\langle\underline{\underline{\tau}}_f\rangle_{LF}\right)\cdot\nabla\alpha_f + (PTI)_f \\
&\quad - (VSTI)_f + \gamma_v\alpha_f{}^t\langle\Gamma_f\rangle^{3i}\langle\underline{U}_f\rangle_{LF} + (MMTI)_f
\end{aligned}$$

$$= -\upsilon^{-1} \int_{A_k} {}^{t}\langle -\nabla_{kf}\sigma_{kf} + 2\sigma_{kf}H_{kf}\underline{n}_k \rangle \, dA$$

$$= -\upsilon^{-1} \int_{A_k} {}^{t}\langle (P_{ck} - P_{cf})\underline{n}_k \rangle \, dA, \qquad (5.6.1)$$

where $P_{ck} - P_{cf}$ is the difference of the capillary pressure. Here, $(PTI)_k$, $(VSTI)_k$, and $(MMTI)_k$ are given in Eqs. (5.5.6c), (5.5.6d), and (5.5.7d), respectively. $(PTI)_f$, $(VSTI)_f$, and $(MMTI)_f$ denote interfacial pressure transfer integral, the interfacial viscous stress transfer integral and interfacial momentum transfer integral for phase f, respectively:

$$(PTI)_f = -\upsilon^{-1} \int_{A_f} \tilde{P}_{fLF}\underline{n}_f dA;$$

$$(VSTI)_f = -\upsilon^{-1} \int_{A_f} \underline{\underline{\tilde{\tau}}}_{fLF} \cdot \underline{n}_f dA \qquad (5.6.2a,b)$$

$$\begin{aligned}
(MMTI)_f = & -\upsilon^{-1} \int_{A_f} [({}^{3i}\langle \rho_f \rangle_{LF} + \tilde{\rho}_{fLF})\underline{\tilde{U}}_{fLF}] \\
& \times ({}^{3i}\langle \underline{U}_f \rangle_{LF} + \underline{\tilde{U}}_{fLF} - \underline{W}_{fLF}) \cdot \underline{n}_f dA \\
& - \upsilon^{-1} \int_{A_f} ({}^{3i}\langle \underline{U}_f \rangle_{LF} + \underline{\tilde{U}}_{fLF} - \underline{W}_{fLF}) \\
& \times {}^{t}\langle \rho'_f \underline{U}'_f \rangle \cdot \underline{n}_f dA \\
& - \upsilon^{-1} \int_{A_f} [({}^{3i}\langle \rho_f \rangle_{LF} + \tilde{\rho}_{fLF})^{t}\langle \underline{U}'_f(\underline{U}'_f - \underline{W}'_f) \rangle \\
& + \underline{\tilde{U}}_{fLF}{}^{t}\langle \rho'_f(\underline{U}'_f - \underline{W}'_f) \rangle] \cdot \underline{n}_f dA \\
& - \upsilon^{-1} \int_{A_f} {}^{t}\langle \rho'_f \underline{U}'_f(\underline{U}'_f - \underline{W}'_f) \rangle \cdot \underline{n}_f dA \qquad (5.6.3)
\end{aligned}$$

$$\begin{aligned}
-\upsilon^{-1} \int_{A_k} {}^{t}\langle P_{ck} - P_{cf} \rangle \underline{n}_k dA = & \gamma_\upsilon {}^{3i}\langle P_{ck} \rangle_{LF} \nabla \alpha_k \\
& - \gamma_\upsilon {}^{3i}\langle P_{cf} \rangle_{LF} \nabla \alpha_f + (CPKI)_k \\
& - (CPFI)_f \qquad (5.6.4)
\end{aligned}$$

$$(CPKI)_k = -v^{-1} \int_{A_k} \tilde{P}_{ckLF} \underline{n}_k dA; \qquad (5.6.5a)$$

$$(CPFI)_f = -v^{-1} \int_{A_f} \tilde{P}_{cfLF} \underline{n}_f dA, \qquad (5.6.5b)$$

where $(CPKI)_k$ and $(CPFI)_f$ denote the interfacial capillary pressure transfer integrals.

It is to be noted that the capillary pressure difference is due to the presence of bubbles, droplets, and so forth in the interfacial momentum balance equation.

5.7 Time-volume-averaged total energy conservation equation

The local volume-averaged total energy conservation equation is given by Eq. (4.6.3):

$$\begin{aligned}
&\gamma_v \frac{\partial}{\partial t} \alpha_k {}^{3i}\langle \rho_k E_k \rangle + \gamma_A \nabla \cdot \alpha_k {}^{2i}\langle \rho_k \underline{U}_k E_k \rangle \\
&= -\gamma_A \nabla \cdot \alpha_k {}^{2i}\langle \underline{U}_k P_k \rangle + \gamma_A \nabla \cdot \alpha_k {}^{2i}\langle \underline{U}_k \cdot \underline{\underline{\tau}}_k \rangle \\
&\quad - \gamma_A \nabla \cdot \alpha_k {}^{2i}\langle \underline{J}_{qk} \rangle + \gamma_v \alpha_k ({}^{3i}\langle \rho_k \underline{U}_k \rangle \cdot \underline{f} \\
&\quad + {}^{3i}\langle J_{Ek} \rangle + {}^{3i}\langle \dot{Q}_{kf} \rangle + {}^{3i}\langle \dot{Q}_{wk} \rangle) \\
&\quad + v^{-1} \int_{A_k} (-P_k \underline{U}_k + \underline{\underline{\tau}}_k \cdot \underline{U}_k) \cdot \underline{n}_k dA \\
&\quad - v^{-1} \int_{A_k} \rho_k E_k (\underline{U}_k - \underline{W}_k) \cdot \underline{n}_k dA, \qquad (4.6.3)
\end{aligned}$$

in which

$$\gamma_v \alpha_k {}^{3i}\langle \dot{Q}_{kf} \rangle = -v^{-1} \int_{A_{kf}} \underline{J}_{qk} \cdot \underline{n}_k dA \qquad (4.6.3a)$$

$$\gamma_v \alpha_k {}^{3i}\langle \dot{Q}_{wk} \rangle = -v^{-1} \int_{A_{wk}} \underline{J}_{qk} \cdot \underline{n}_k dA, \qquad (4.6.3b)$$

where $\gamma_v \alpha_k {}^{3i}\langle \dot{Q}_{kf} \rangle$ denotes the fluid-to-fluid interfacial heat transfer rate into phase k per unit volume of phase k in v, and $\gamma_v \alpha_k {}^{3i}\langle \dot{Q}_{wk} \rangle$ denotes the fluid-to-solid structure interfacial heat transfer rate into phase k per unit volume of phase k in v.

For the total energy, we may also write

$$E_k = {}^{3i}\langle E_k \rangle_{LF} + \tilde{E}_{kLF} + E'_k. \tag{5.7.1}$$

However, the defining relations for ${}^{3i}\langle E_k \rangle_{LF}$, $\tilde{E}_{kLF}$, and E'_k require careful consideration. Specifically, ${}^{3i}\langle E_k \rangle_{LF}$ is of low frequency and of length scale l; $\tilde{E}_{kLF}$ is also of low frequency and must satisfy the condition ${}^{3i}\langle \tilde{E}_{kLF} \rangle = 0$; and the time average of E'_k should vanish. Because $E_k = u_k + \frac{1}{2}\underline{U}_k \cdot \underline{U}_k$, and by using the already defined relations $u_k = {}^{3i}\langle u_k \rangle_{LF} + \tilde{u}_{kLF} + u'_k$ and $\underline{U}_k = {}^{3i}\langle \underline{U}_k \rangle_{LF} + \underline{\tilde{U}}_{kLF} + \underline{U}'_k$, we obtain the following:

$$\begin{aligned} E_k = {}& {}^{3i}\langle u_k \rangle_{LF} + \tilde{u}_{kLF} + u'_k + \frac{1}{2} {}^{3i}\langle \underline{U}_k \rangle_{LF} \cdot {}^{3i}\langle \underline{U}_k \rangle_{LF} \\ & + \frac{1}{2}\underline{\tilde{U}}_{kLF} \cdot \underline{\tilde{U}}_{kLF} + \frac{1}{2}\underline{U}'_k \cdot \underline{U}'_k + {}^{3i}\langle \underline{U}_k \rangle_{LF} \cdot \underline{\tilde{U}}_{kLF} \\ & + {}^{3i}\langle \underline{U}_k \rangle_{LF} \cdot \underline{U}'_k + \underline{\tilde{U}}_{kLF} \cdot \underline{U}'_k. \end{aligned} \tag{5.7.1a}$$

We define

$$\begin{aligned} {}^{3i}\langle E_k \rangle_{LF} = {}& {}^{3i}\langle u_k \rangle_{LF} + \frac{1}{2} {}^{3i}\langle \underline{U}_k \rangle_{LF} \cdot {}^{3i}\langle \underline{U}_k \rangle_{LF} \\ & + \frac{1}{2} {}^{3i}\langle \underline{\tilde{U}}_{kLF} \cdot \underline{\tilde{U}}_{kLF} \rangle \end{aligned} \tag{5.7.1b}$$

$$\begin{aligned} \tilde{E}_{kLF} = {}& \tilde{u}_{kLF} + {}^{3i}\langle \underline{U}_k \rangle_{LF} \cdot \underline{\tilde{U}}_{kLF} \\ & + \frac{1}{2}(\underline{\tilde{U}}_{kLF} \cdot \underline{\tilde{U}}_{kLF} - {}^{3i}\langle \underline{\tilde{U}}_{kLF} \cdot \underline{\tilde{U}}_{kLF} \rangle) \end{aligned} \tag{5.7.1c}$$

$$E'_k = u'_k + {}^{3i}\langle \underline{U}_k \rangle_{LF} \cdot \underline{U}'_k + \underline{\tilde{U}}_{kLF} \cdot \underline{U}'_k$$
$$+ \frac{1}{2}(\underline{U}'_k \cdot \underline{U}'_k - {}^t\langle \underline{U}'_k \cdot \underline{U}'_k \rangle). \quad (5.7.1d)$$

We can readily verify that Eq. (5.7.1a) with the defining relations for ${}^{3i}\langle E_k \rangle_{LF}$, $\tilde{E}_{kLF}$, and E'_k, given by Eqs. (5.7.1b), (5.7.1c), and (5.7.1d), respectively, and their sum is identical to Eq. (5.7.1a). Furthermore, the constraint just cited for ${}^{3i}\langle \tilde{E}_{kLF} \rangle = 0$ is satisfied, and other constraint ${}^t\langle E'_k \rangle = 0$ is also satisfied since $\frac{1}{2}{}^t\langle \underline{U}'_k \cdot \underline{U}'_k \rangle \ll {}^{3i}\langle \underline{U}_k \rangle_{LF} \cdot \underline{U}'_k$ and $\frac{1}{2}{}^t\langle \underline{U}'_k \cdot \underline{U}'_k \rangle$ can be neglected.

Time averaging of the local volume-averaged total energy conservation equations requires consideration of the following.

The first term on the LHS of Eq. (4.6.3) is

$$\gamma_v \frac{\partial}{\partial t} {}^t\langle \alpha_k {}^{3i}\langle \rho_k E_k \rangle\rangle \quad (5.7.2)$$

$${}^t\langle \alpha_k {}^{3i}\langle \rho_k E_k \rangle\rangle = \alpha_k {}^{3i}\langle \rho_k \rangle_{LF} {}^{3i}\langle E_k \rangle_{LF} + \phi^{3i}_{Ek}, \quad (5.7.2a)$$

in which ϕ^{3i}_{EK} is a scalar total energy function defined by

$$\phi^{3i}_{Ek} = \alpha_k {}^{3i}\langle \tilde{\rho}_{kLF} \tilde{E}_{kLF} \rangle + \alpha_k {}^{t3i}\langle \rho'_k E'_k \rangle. \quad (5.7.2b)$$

The first term on the LHS of Eq. (4.6.3) can be written as follows:

$$\gamma_v \frac{\partial}{\partial t} {}^t\langle \alpha_k {}^{3i}\langle \rho_k E_k \rangle\rangle = \gamma_v \frac{\partial}{\partial t} \alpha_k {}^{3i}\langle \rho_k \rangle_{LF} {}^{3i}\langle E_k \rangle_{LF}$$
$$+ \gamma_v \frac{\partial \phi^{3i}_{Ek}}{\partial t}. \quad (5.7.2c)$$

The second term on the LHS of Eq. (4.6.3) is

$$\gamma_A \nabla \cdot {}^{t}\langle \alpha_k {}^{2i}\langle \rho_k \underline{U}_k E_k \rangle\rangle \tag{5.7.3}$$

$$\begin{aligned}
{}^{t}\langle \alpha_k {}^{2i}\langle \rho_k \underline{U}_k E_k \rangle\rangle &= {}^{t}\langle \alpha_k ({}^{2i}\langle \rho_k \rangle_{LF} + \tilde{\rho}_{kLF} + \rho'_k)({}^{2i}\langle \underline{U}_k \rangle_{LF} \\
&\quad + \underline{\tilde{U}}_{kLF} + \underline{U}'_k)({}^{2i}\langle E_k \rangle_{LF} + \tilde{E}_{kLF} + E'_k)\rangle \\
&= \alpha_k {}^{2i}\langle \rho_k \rangle_{LF} {}^{2i}\langle \underline{U}_k \rangle_{LF} {}^{2i}\langle E_k \rangle_{LF} \\
&\quad + {}^{2i}\langle \underline{U}_k \rangle_{LF} \phi^{2i}_{Ek} + {}^{2i}\langle E_k \rangle_{LF} \underline{\psi}^{2i}_{mk} \\
&\quad + \alpha_k \left({}^{2i}\langle \underline{E}^T_k \rangle + {}^{2i}\langle \underline{\tilde{E}}_k \rangle + {}^{2i}\langle \underline{\tilde{E}}^T_k \rangle \right),
\end{aligned} \tag{5.7.3a}$$

where

$$\phi^{2i}_{Ek} = \alpha_k {}^{2i}\langle \tilde{\rho}_{kLF} \tilde{E}_{kLF} \rangle + \alpha_k {}^{t2i}\langle \rho'_k E'_k \rangle. \tag{5.7.3b}$$

$\underline{\psi}^{2i}_{mk}$ is defined as in Eq. (5.3.2h).

(a) The volume-averaged turbulent total energy flux vector ${}^{2i}\langle \underline{E}^T_k \rangle$ is defined by

$${}^{2i}\langle \rho_k \rangle_{LF} {}^{t2i}\langle \underline{U}'_k E'_k \rangle + {}^{t2i}\langle \tilde{\rho}_{kLF} \underline{U}'_k E'_k \rangle. \tag{5.7.3c}$$

(b) The volume-averaged dispersive total energy flux vector ${}^{2i}\langle \underline{\tilde{E}}_k \rangle$ is defined by

$${}^{2i}\langle \rho_k \rangle_{LF} {}^{2i}\langle \underline{\tilde{U}}_{kLF} \tilde{E}_{kLF} \rangle + {}^{2i}\langle \tilde{\rho}_{kLF} \underline{\tilde{U}}_{kLF} \tilde{E}_{kLF} \rangle. \tag{5.7.3d}$$

(c) The volume-averaged turbulent, dispersive total energy flux vector ${}^{2i}\langle \underline{\tilde{E}}^T_k \rangle$ is defined by

$${}^{t2i}\langle \underline{\tilde{U}}_{kLF} \rho'_k E'_k \rangle + {}^{t2i}\langle \tilde{E}_{kLF} \rho'_k \underline{U}'_k \rangle. \tag{5.7.3e}$$

(d) $\rho'_k \underline{U}'_k E'_k$ is a triple time correlation and assumed to be small; thus, it can be neglected. (5.7.3f)

The second term on the LHS of Eq. (4.6.3) can be written as follows:

$$\gamma_A \nabla \cdot {}^t\langle \alpha_k {}^{2i}\langle \rho_k \underline{U}_k E_k \rangle\rangle$$

$$= \gamma_A \nabla \cdot \big[\alpha_k {}^{2i}\langle \rho_k\rangle_{LF} {}^{2i}\langle \underline{U}_k\rangle_{LF} {}^{2i}\langle E_k\rangle_{LF}$$

$$+ {}^{2i}\langle \underline{U}_k\rangle_{LF} \phi^{2i}_{Ek} + {}^{2i}\langle E_k\rangle_{LF} \underline{\psi}^{2i}_{mk}$$

$$+ \alpha_k \big({}^{2i}\langle \underline{E}^T_k\rangle + {}^{2i}\langle \underline{\tilde{E}}_k\rangle + {}^{2i}\langle \underline{\tilde{E}}^T_k\rangle\big)\big]. \quad (5.7.3g)$$

The first term on the RHS of Eq. (4.6.3) is

$$-\gamma_A \nabla \cdot {}^t\langle \alpha_k {}^{2i}\langle \underline{U}_k P_k\rangle\rangle \quad (5.7.4)$$

$${}^t\langle -\alpha_k {}^{2i}\langle \underline{U}_k P_k\rangle\rangle$$

$$= {}^t\langle {}^{2i}\langle \alpha_k ({}^{2i}\langle \underline{U}_k\rangle_{LF} + \underline{\tilde{U}}_{kLF} + \underline{U}'_k)({}^{2i}\langle P_k\rangle_{LF} + \tilde{P}_{kLF} + P'_k)\rangle\rangle$$

$$= -\alpha_k {}^{2i}\langle \underline{U}_k\rangle_{LF} {}^{2i}\langle P_k\rangle_{LF} - \underline{\Psi}^{2i}_{Pk}, \quad (5.7.4a)$$

where $\underline{\Psi}^{2i}_{Pk}$ is a pressure work function vector defined by

$$\underline{\Psi}^{2i}_{Pk} = \alpha_k {}^{2i}\langle \underline{\tilde{U}}_{kLF} \tilde{P}_{kLF}\rangle + \alpha_k {}^{t2i}\langle \underline{U}'_k P'_k\rangle. \quad (5.7.4b)$$

The first term on the RHS of Eq. (4.6.3) can be written as follows:

$$-\gamma_A \nabla \cdot {}^{2i}\langle \alpha_k {}^{2i}\langle \underline{U}_k P_k\rangle\rangle$$

$$= -\gamma_A \nabla \cdot (\alpha_k {}^{2i}\langle \underline{U}_k\rangle_{LF} {}^{2i}\langle P_k\rangle_{LF} + \underline{\Psi}^{2i}_{Pk}). \quad (5.7.4c)$$

The second term on the RHS of Eq. (4.6.3) is

$$-\gamma_A \nabla \cdot {}^t\langle \alpha_k {}^{2i}\langle \underline{U}_k \cdot \underline{\underline{\tau}}_k \rangle\rangle \tag{5.7.5}$$

$$\begin{aligned} {}^t\langle \alpha_k {}^{2i}\langle \underline{U}_k \cdot \underline{\underline{\tau}}_k \rangle\rangle &= {}^t\langle \alpha_k {}^{2i}\langle ({}^{2i}\langle \underline{U}_k \rangle_{LF} + \underline{\tilde{U}}_{kLF} + \underline{U}'_k) \\ &\quad \cdot ({}^{2i}\langle \underline{\underline{\tau}}_k \rangle_{LF} + \underline{\underline{\tilde{\tau}}}_{kLF} + \underline{\underline{\tau}}'_k)\rangle\rangle \\ &= \alpha_k {}^{2i}\langle \underline{U}_k \rangle_{LF} \cdot {}^{2i}\langle \underline{\underline{\tau}}_k \rangle_{LF} + \underline{\Psi}^{2i}_{\tau k} \end{aligned} \tag{5.7.5a}$$

Here, ${}^{2i}\langle \underline{\underline{\tau}}_k \rangle_{LF}$, $\underline{\underline{\tilde{\tau}}}_{kLF}$ and $\underline{\underline{\tau}}'_k$ are defined as in Eqs. (5.5.4f), (5.5.4g), and (5.5.4h), respectively, and $\underline{\Psi}^{2i}_{\tau k}$ is a viscous stress work function vector defined by

$$\underline{\Psi}^{2i}_{\tau k} = \alpha_k {}^{2i}\langle \underline{\tilde{U}}_{kLF} \cdot \underline{\underline{\tilde{\tau}}}_{kLF} \rangle + \alpha_k {}^{t2i}\langle \underline{U}'_k \cdot \underline{\underline{\tau}}'_k \rangle. \tag{5.7.5b}$$

The second term on the RHS of Eq. (4.6.3) can be written as follows:

$$\begin{aligned} &\gamma_A \nabla \cdot {}^t\langle \alpha_k {}^{2i}\langle \underline{U}_k \cdot \underline{\underline{\tau}}_k \rangle\rangle \\ &= \gamma_A \nabla \cdot (\alpha_k {}^{2i}\langle \underline{U}_k \rangle_{LF} \cdot {}^{2i}\langle \underline{\underline{\tau}}_k \rangle_{LF} + \underline{\Psi}^{2i}_{\tau k}). \end{aligned} \tag{5.7.5c}$$

The third term on the RHS of Eq. (4.6.3) is

$$-\gamma_A \nabla \cdot {}^t\langle \alpha_k {}^{2i}\langle \underline{J}_{qk} \rangle\rangle \tag{5.7.6}$$

$$\begin{aligned} {}^t\langle \alpha_k {}^{2i}\langle \underline{J}_{qk} \rangle\rangle &= {}^t\langle \alpha_k {}^{2i}\langle {}^{2i}\langle \underline{J}_{qk} \rangle_{LF} + \underline{\tilde{J}}_{qkLF} + \underline{J}'_{qk} \rangle\rangle \\ &= \alpha_k {}^{2i}\langle \underline{J}_{qk} \rangle_{LF}. \end{aligned} \tag{5.7.6a}$$

The Fourier law of isotropic conduction states that

$$\underline{J}_{qk} = -\kappa_k \nabla T_k. \tag{5.7.6b}$$

If the thermal conductivity κ is independent of T_k, then ${}^{2i}\langle \underline{J}_{qk} \rangle_{LF} = -\kappa_k \nabla {}^{2i}\langle T_k \rangle_{LF}$.

When expressed in terms of internal energy u_k, Eq. (5.7.6b) becomes, for constant specific heat c_{vk}, as follows:

$$ {}^{2i}\langle \underline{J}_{qk} \rangle_{LF} = -\frac{\kappa_k}{c_{vk}} \nabla\, {}^{2i}\langle u_k \rangle_{LF}. \tag{5.7.6c} $$

The case of variable conductivity and specific heat is treated in Appendix D.

The third term on the RHS of Eq. (4.6.3) can be written as follows:

$$ \begin{aligned} -\gamma_A \nabla \cdot {}^{t}\langle \alpha_k\, {}^{2i}\langle \underline{J}_{qk} \rangle \rangle &= -\gamma_A \nabla \cdot \alpha_k\, {}^{2i}\langle \underline{J}_{qk} \rangle_{LF} \\ &= \gamma_A \nabla \cdot \left(\alpha_k \frac{\kappa_k}{c_{vk}} \nabla\, {}^{2i}\langle u_k \rangle_{LF} \right). \end{aligned} \tag{5.7.6d} $$

The fourth term on the RHS of Eq. (4.6.3) is

$$ \gamma_v\, {}^{t}\langle \alpha_k ({}^{3i}\langle \rho_k \underline{U}_k \rangle \cdot \underline{f} + {}^{3i}\langle J_{Ek} \rangle + {}^{3i}\langle \dot{Q}_{kf} \rangle + {}^{3i}\langle \dot{Q}_{wk} \rangle) \rangle \tag{5.7.7} $$

$$ \begin{aligned} & {}^{t}\langle \alpha_k\, {}^{3i}\langle \rho_k \underline{U}_k \rangle \cdot \underline{f} \rangle \\ &= {}^{t}\langle \alpha_k\, {}^{3i}\langle ({}^{3i}\langle \rho_k \rangle_{LF} + \tilde{\rho}_{kLF} + \rho'_k) \\ &\quad \times ({}^{3i}\langle \underline{U}_k \rangle_{LF} + \underline{\tilde{U}}_{kLF} + \underline{U}'_k) \cdot \underline{f} \rangle \rangle \\ &= (\alpha_k\, {}^{3i}\langle \rho_k \rangle_{LF}\, {}^{3i}\langle \underline{U}_k \rangle_{LF} \cdot \underline{f} + \underline{\psi}^{3i}_{mk} \cdot \underline{f}). \end{aligned} \tag{5.7.7a} $$

Here, $\underline{\psi}^{3i}_{mk}$ is the mass flux vector and is defined in Eq. (5.5.1b):

$$ \begin{aligned} \gamma_v\, {}^{t}\langle \alpha_k\, {}^{3i}\langle J_{Ek} \rangle \rangle &= \gamma_v\, {}^{t}\langle \alpha_k\, {}^{3i}\langle {}^{3i}\langle J_{Ek} \rangle_{LF} + \tilde{J}_{EkLF} + J'_{Ek} \rangle \rangle \\ &= \gamma_v \alpha_k\, {}^{3i}\langle J_{Ek} \rangle_{LF} \end{aligned} \tag{5.7.7b} $$

$$
\begin{aligned}
&{}^{t}\left\langle -\upsilon^{-1}\int_{A_{kf}} \underline{J}_{qk}\cdot\underline{n}_k dA\right\rangle \\
&= -\upsilon^{-1}\int_{A_{kf}} {}^{t}\langle\underline{J}_{qk}\rangle\cdot\underline{n}_k dA \\
&= \gamma_\upsilon{}^{3i}\langle\underline{J}_{qk}\rangle_{LF}\cdot\nabla\alpha_k - \upsilon^{-1}\int_{A_{kf}} \tilde{\underline{J}}_{qkLF}\cdot\underline{n}_k dA \\
&= \gamma_\upsilon{}^{3i}\langle\underline{J}_{qk}\rangle_{LF}\cdot\nabla\alpha_k + (HTI)_k \\
&= \gamma_\upsilon\alpha_k{}^{t3i}\langle\dot{Q}_{kf}\rangle, \qquad (5.7.7c)
\end{aligned}
$$

where $(HTI)_k$ is the interfacial heat transfer integral defined as follows:

$$
(HTI)_k = -\upsilon^{-1}\int_{A_{kf}} \tilde{\underline{J}}_{qkLF}\cdot\underline{n}_k dA \qquad (5.7.7d)
$$

$$
-\upsilon^{-1}\int_{wk} {}^{t}\langle\underline{J}_{qk}\cdot\underline{n}_k dA\rangle = \gamma_\upsilon\alpha_k{}^{t3i}\langle\dot{Q}_{wk}\rangle. \qquad (5.7.7e)
$$

Here, $\gamma_\upsilon\alpha_k{}^{t3i}\langle\dot{Q}_{wk}\rangle$ represents part of the distributed heat source and sink due to the presence of heat-generating and heat-absorbing internal stationary structures.

The fourth term on the RHS of Eq. (4.6.3) can be written as follows:

$$
\begin{aligned}
&\gamma_\upsilon{}^{t}\langle\alpha_k({}^{3i}\langle\rho_k\underline{U}_k\rangle\cdot\underline{f} + {}^{3i}\langle J_{Ek}\rangle + {}^{3i}\langle\dot{Q}_{kf}\rangle + {}^{3i}\langle\dot{Q}_{wk}\rangle)\rangle \\
&= \gamma_\upsilon[(\alpha_k{}^{3i}\langle\rho_k\rangle_{LF}{}^{3i}\langle\underline{U}_k\rangle_{LF}\cdot\underline{f} + \underline{\psi}^{3i}_{mk}\cdot\underline{f}) \\
&\quad + \alpha_k{}^{3i}\langle\underline{J}_{Ek}\rangle_{LF} + \alpha_k{}^{t3i}\langle\dot{Q}_{kf}\rangle + \alpha_k{}^{t3i}\langle\dot{Q}_{wk}\rangle]. \qquad (5.7.7f)
\end{aligned}
$$

The fifth term on the RHS of Eq. (4.6.3) is

$$
\begin{aligned}
&{}^{t}\left\langle \upsilon^{-1}\int_{A_k} (-\underline{U}_k P_k + \underline{U}_k\cdot\underline{\underline{\tau}}_k)\cdot\underline{n}_k dA\right\rangle \\
&= \upsilon^{-1}\int_{A_k} {}^{t}\langle(-\underline{U}_k P_k + \underline{U}_k\cdot\underline{\underline{\tau}}_k)\cdot\underline{n}_k dA\rangle \qquad (5.7.8)
\end{aligned}
$$

$$
\begin{aligned}
&\upsilon^{-1}\int_{A_k} {}^{t}\langle -\underline{U}_k P_k + \underline{U}_k \cdot \underline{\underline{\tau}}_k\rangle \cdot \underline{n}_k dA \\
&= \gamma_\upsilon\, {}^{3i}\langle \underline{U}_k\rangle_{LF}\, {}^{3i}\langle P_k\rangle_{LF} \cdot \nabla\alpha_k \\
&- \upsilon^{-1}\, {}^{3i}\langle P_k\rangle_{LF} \int_{A_k} \underline{\tilde{U}}_{kLF} \cdot \underline{n}_k dA \\
&- \upsilon^{-1}\, {}^{3i}\langle \underline{U}_k\rangle_{LF} \int_{A_k} \tilde{P}_{kLF} \underline{n}_k dA \\
&- \upsilon^{-1} \int_{A_k} \underline{\tilde{U}}_{kLF} \tilde{P}_{kLF} \cdot \underline{n}_k dA - \upsilon^{-1} \int_{A_k} {}^{t}\langle \underline{U}'_k P'_k\rangle \cdot \underline{n}_k dA \\
&- \gamma_\upsilon\, {}^{3i}\langle \underline{U}_k\rangle_{LF} \cdot {}^{3i}\langle \underline{\underline{\tau}}_k\rangle_{LF} \cdot \nabla\alpha_k \\
&+ \upsilon^{-1} \int_{A_k} \left(\underline{\tilde{U}}_{kLF} \cdot {}^{3i}\langle \underline{\underline{\tau}}_k\rangle_{LF}\right) \cdot \underline{n}_k dA \\
&+ \upsilon^{-1} \int_{A_k} \left({}^{3i}\langle \underline{U}_k\rangle_{LF} \cdot \underline{\underline{\tilde{\tau}}}_{kLF}\right) \cdot \underline{n}_k dA \\
&+ \upsilon^{-1} \int_{A_k} \left(\underline{\tilde{U}}_{kLF} \cdot \underline{\underline{\tilde{\tau}}}_{kLF}\right) \cdot \underline{n}_k dA \\
&+ \upsilon^{-1} \int_{A_k} {}^{t}\langle \underline{U}'_k \cdot \underline{\underline{\tau}}'_k\rangle \cdot \underline{n}_k dA, \qquad (5.7.8a)
\end{aligned}
$$

in which the interfacial pressure work integral $(PWI)_k$ is defined by

$$
\begin{aligned}
(PWI)_k = &-\upsilon^{-1}\, {}^{3i}\langle P_k\rangle_{LF} \int_{A_k} \underline{\tilde{U}}_{kLF} \cdot \underline{n}_k dA \\
&-\upsilon^{-1}\, {}^{3i}\langle \underline{U}_k\rangle_{LF} \int_{A_k} \tilde{P}_{kLF} \cdot \underline{n}_k dA \\
&-\upsilon^{-1} \int_{A_k} \left(\underline{\tilde{U}}_{kLF} \tilde{P}_{kLF} + {}^{t}\langle \underline{U}'_k P'_k\rangle\right) \cdot \underline{n}_k dA. \qquad (5.7.8b)
\end{aligned}
$$

The interfacial viscous stress work integral $(VWI)_k$ is defined as follows:

$$(VWI)_k = -\upsilon^{-1}\int_{A_k} (\underline{\tilde{U}}_{kLF} \cdot {}^{3i}\langle\underline{\underline{\tau}}_k\rangle_{LF}) \cdot \underline{n}_k dA$$
$$-\upsilon^{-1}\int_{A_k} ({}^{3i}\langle\underline{U}_k\rangle_{LF} \cdot \underline{\underline{\tilde{\tau}}}_{kLF}) \cdot \underline{n}_k dA$$
$$-\upsilon^{-1}\int_{A_k} (\underline{\tilde{U}}_{kLF} \cdot \underline{\underline{\tilde{\tau}}}_{kLF} + {}^{t}\langle\underline{U}'_k \cdot \underline{\underline{\tau}}'_k\rangle) \cdot \underline{n}_k dA. \tag{5.7.8c}$$

The fifth term on the RHS of Eq. (4.6.3) can be written as follows:

$$\upsilon^{-1}\int_{A_k} {}^{t}\langle(-\underline{U}_k P_k + \underline{U}_k \cdot \underline{\underline{\tau}}_k) \cdot \underline{n}_k dA\rangle$$
$$= \gamma_\upsilon {}^{3i}\langle\underline{U}_k\rangle_{LF} {}^{3i}\langle P_k\rangle_{LF} \cdot \nabla\alpha_k - \gamma_\upsilon({}^{3i}\langle\underline{U}_k\rangle_{LF} \cdot {}^{3i}\langle\underline{\underline{\tau}}_k\rangle_{LF}) \cdot \nabla\alpha_k$$
$$+ (PWI)_k - (VWI)_k. \tag{5.7.8d}$$

The sixth term on the RHS of Eq. (4.6.3) is

$${}^{t}\left\langle -\upsilon^{-1}\int_{A_k} \rho_k E_k (\underline{U}_k - \underline{W}_k) \cdot \underline{n}_k dA \right\rangle$$
$$= -\upsilon^{-1}\int_{A_k} {}^{t}\langle \rho_k E_k (\underline{U}_k - \underline{W}_k) \cdot \underline{n}_k dA\rangle \tag{5.7.9}$$

$$-\upsilon^{-1}\int_{A_k} {}^{t}\langle \rho_k E_k (\underline{U}_k - \underline{W}_k) \cdot \underline{n}_k dA\rangle$$
$$= -\upsilon^{-1}\int_{A_k} {}^{t}\langle({}^{3i}\langle\rho_k\rangle_{LF} + \tilde{\rho}_{kLF} + \rho'_k)$$
$$\times ({}^{3i}\langle E_k\rangle_{LF} + \tilde{E}_{kLF} + E'_k)$$
$$\times [({}^{3i}\langle\underline{U}_k\rangle_{LF} + \underline{\tilde{U}}_{kLF} + \underline{U}'_k) - (\underline{W}_{kLF} + \underline{W}'_k)] \cdot \underline{n}_k dA. \tag{5.7.9a}$$

Because low- and high-frequency parts of Eq. (5.7.9a) can be separated, after the time averaging the equation becomes

$$
\begin{aligned}
& \upsilon^{-1} \int_{A_k} {}^{t}\langle \rho_k E_k (\underline{U}_k - \underline{W}_k) \cdot \underline{n}_k dA \rangle \\
& = \gamma_\upsilon {}^{3i}\langle \rho_k \rangle_{LF} {}^{3i}\langle E_k \rangle_{LF} \left(\frac{\partial \alpha_k}{\partial t} + {}^{3i}\langle \underline{U}_k \rangle_{LF} \cdot \nabla \alpha_k \right) \\
& - \upsilon^{-1} {}^{3i}\langle \rho_k \rangle_{LF} {}^{3i}\langle E_k \rangle_{LF} \int_{A_k} \underline{\tilde{U}}_{kLF} \cdot \underline{n}_k dA \\
& - \upsilon^{-1} {}^{3i}\langle E_k \rangle_{LF} \int_{A_k} \tilde{\rho}_{kLF} ({}^{3i}\langle \underline{U}_k \rangle_{LF} + \underline{\tilde{U}}_{kLF} - \underline{W}_{kLF}) \cdot \underline{n}_k dA \\
& - \upsilon^{-1} \int_{A_k} [({}^{3i}\langle \rho_k \rangle_{LF} + \tilde{\rho}_{kLF}) \tilde{E}_{kLF}] \\
& \times ({}^{3i}\langle \underline{U}_k \rangle_{LF} + \underline{\tilde{U}}_{kLF} - \underline{W}_{kLF}) \cdot \underline{n}_k dA \\
& - \upsilon^{-1} \int_{A_k} ({}^{3i}\langle \rho_k \rangle_{LF} + \tilde{\rho}_{kLF}) {}^{t}\langle E_k' (\underline{U}_k' - \underline{W}_k') \rangle \cdot \underline{n}_k dA \\
& - \upsilon^{-1} \int_{A_k} ({}^{3i}\langle E_k \rangle_{LF} + \tilde{E}_{kLF}) {}^{t}\langle \rho_k' (\underline{U}_k' - \underline{W}_k') \rangle \cdot \underline{n}_k dA \\
& - \upsilon^{-1} \int_{A_k} ({}^{3i}\langle \underline{U}_k \rangle_{LF} + \underline{\tilde{U}}_{kLF} - \underline{W}_{kLF}) {}^{t}\langle \rho_k' E_k' \rangle \cdot \underline{n}_k dA \\
& - \upsilon^{-1} \int_{A_k} {}^{t}\langle \rho_k' E_k' (\underline{U}_k' - \underline{W}_k') \cdot \underline{n}_k dA \rangle. \qquad (5.7.9b)
\end{aligned}
$$

The sixth term on the RHS of Eq. (4.6.3) can be written by multiplying Eq. (5.3.3c) by ${}^{3i}\langle E_k \rangle_{LF}$. When the results are introduced into Eq. (5.7.9b), we obtain

$$
\begin{aligned}
& -\upsilon^{-1} \int_{A_k} {}^{t}\langle \rho_k E_k (\underline{U}_k - \underline{W}_k) \rangle \cdot \underline{n}_k dA \\
& = \gamma_\upsilon \alpha_k {}^{t}\langle \Gamma_k \rangle {}^{3i}\langle E_k \rangle_{LF} + (TETI)_k . \qquad (5.7.9c)
\end{aligned}
$$

The interfacial total energy transfer integral is

$$
\begin{aligned}
(TETI)_k \\
= -\upsilon^{-1} \int_{A_k} & \tilde{E}_{kLF}({}^{3i}\langle \rho_k \rangle_{LF} + \tilde{\rho}_{kLF}) \\
& \times ({}^{3i}\langle \underline{U}_k \rangle_{LF} + \underline{\tilde{U}}_{kLF} - \underline{W}_{kLF}) \cdot \underline{n}_k dA \\
- \upsilon^{-1} \int_{A_k} & ({}^{3i}\langle \rho_k \rangle_{LF} + \tilde{\rho}_{kLF})\, {}^{t}\langle E'_k(\underline{U}'_k - \underline{W}'_k) \rangle \cdot \underline{n}_k dA \\
- \upsilon^{-1} \int_{A_k} & \tilde{E}_{kLF}\, {}^{t}\langle \rho'_k(\underline{U}'_k - \underline{W}'_k) \rangle \cdot \underline{n}_k dA \\
- \upsilon^{-1} \int_{A_k} & ({}^{3i}\langle \underline{U}_k \rangle_{LF} + \underline{\tilde{U}}_{kLF} - \underline{W}_{kLF})\, {}^{t}\langle \rho'_k E'_k \rangle \cdot \underline{n}_k dA \\
- \upsilon^{-1} \int_{A_k} & {}^{t}\langle \rho'_k E'_k (\underline{U}'_k - \underline{W}'_k) \rangle \cdot \underline{n}_k dA. \qquad (5.7.9d)
\end{aligned}
$$

Once again, we see that the time-volume-averaged interfacial total energy transfer rate consists of three parts: (1) interfacial heat transfer $\gamma_\upsilon \alpha_k ({}^{t3i}\langle \dot{Q}_{kf} \rangle + {}^{t3i}\langle \dot{Q}_{wk} \rangle)$, as given by Eqs. (5.7.7c) and (5.7.7e); (2) interfacial work done by pressure force and viscous force, as given by Eq. (5.7.8d), and (3) interfacial total energy transfer due to interfacial mass generation plus interfacial energy transfer integrals, as given by Eq. (5.7.9c). The time-averaged interfacial total energy source of phase k per unit volume in υ, $\gamma_\upsilon \alpha_k {}^{t}\langle \Xi_k \rangle$, resulting from the interfacial total energy transfer, is

$$
\begin{aligned}
\gamma_\upsilon \alpha_k {}^{t}\langle \Xi_k \rangle = {} & \gamma_\upsilon \alpha_k ({}^{t3i}\langle \dot{Q}_{kf} \rangle + {}^{t3i}\langle \dot{Q}_{wk} \rangle) \\
& + \gamma_\upsilon {}^{3i}\langle \underline{U}_k \rangle_{LF} {}^{3i}\langle P_k \rangle_{LF} \cdot \nabla \alpha_k \\
& - \gamma_\upsilon {}^{3i}(\langle \underline{U}_k \rangle_{LF} \cdot {}^{3i}\langle \underline{\underline{\tau}}_k \rangle_{LF}) \cdot \nabla \alpha_k \\
& + (PWI)_k - (VWI)_k \\
& + \gamma_\upsilon \alpha_k {}^{t}\langle \Gamma_k \rangle {}^{3i}\langle E_k \rangle_{LF} + (TETI)_k. \qquad (5.7.9e)
\end{aligned}
$$

Using the results given in Eqs. (5.7.2c), (5.7.3g), (5.7.4c), (5.7.5c), (5.7.6d), (5.7.7f), (5.7.8d), and (5.7.9c), we obtain the time-volume-averaged total energy conservation equation:

$$\begin{aligned}
&\gamma_v \frac{\partial}{\partial t}(\alpha_k {}^{3i}\langle\rho_k\rangle_{LF} {}^{3i}\langle E_k\rangle_{LF} + \phi_{Ek}^{3i}) \\
&\quad + \gamma_A \nabla \cdot [\alpha_k {}^{2i}\langle\rho_k\rangle_{LF} {}^{2i}\langle\underline{U}_k\rangle_{LF} {}^{2i}\langle E_k\rangle_{LF} + \phi_{Ek}^{2i} {}^{2i}\langle\underline{U}_k\rangle_{LF} \\
&\quad + \underline{\psi}_{mk}^{2i} {}^{2i}\langle E_k\rangle_{LF} + \alpha_k({}^{2i}\langle\underline{E}_k^T\rangle + {}^{2i}\langle\underline{\tilde{E}}_k\rangle + {}^{2i}\langle\underline{\tilde{E}}_k^T\rangle)] \\
&= -\gamma_A \nabla \cdot (\alpha_k {}^{2i}\langle\underline{U}_k\rangle_{LF} {}^{2i}\langle P_k\rangle_{LF} + \underline{\Psi}_{Pk}^{2i}) \\
&\quad + \gamma_A \nabla \cdot (\alpha_k {}^{2i}\langle\underline{U}_k\rangle_{LF} \cdot {}^{2i}\langle\underline{\underline{\tau}}_k\rangle_{LF} + \underline{\Psi}_{\tau k}^{2i}) \\
&\quad - \gamma_A \nabla \cdot \alpha_k {}^{2i}\langle\underline{J}_{qk}\rangle_{LF} + \gamma_v[(\alpha_k {}^{3i}\langle\rho_k\rangle_{LF} {}^{3i}\langle\underline{U}_k\rangle_{LF} \cdot \underline{f} \\
&\quad + \underline{\psi}_{mk}^{3i} \cdot \underline{f}) + \alpha_k {}^{3i}\langle J_{Ek}\rangle_{LF} + \alpha_k {}^{t3i}\langle\dot{Q}_{kf}\rangle + \alpha_k {}^{t3i}\langle\dot{Q}_{wk}\rangle] \\
&\quad + \gamma_v {}^{3i}\langle\underline{U}_k\rangle_{LF} {}^{3i}\langle P_k\rangle_{LF} \cdot \nabla\alpha_k \\
&\quad - \gamma_v ({}^{3i}\langle\underline{U}_k\rangle_{LF} \cdot {}^{3i}\langle\underline{\underline{\tau}}_k\rangle_{LF}) \cdot \nabla\alpha_k + (PWI)_k - (VWI)_k \\
&\quad + \gamma_v \alpha_k {}^{t}\langle\Gamma_k\rangle {}^{3i}\langle E_k\rangle_{LF} + (TETI)_k . \qquad (5.7.10)
\end{aligned}$$

When $\rho_k =$ constant, ${}^{3i}\langle\rho_k\rangle_{LF} = \rho_k$, $\tilde{\rho}_{kLF} = \rho_k' = 0$. Therefore,

$$\phi_{Ek}^{3i} = 0, \quad \phi_{Ek}^{2i} = 0, \quad \underline{\psi}_{mk}^{3i} = 0 \quad \text{and} \quad \underline{\psi}_{mk}^{2i} = 0 \quad (5.7.11)$$

$${}^{o2i}\langle\underline{E}_k^T\rangle = \rho_k \underline{U}_k' E_k' \qquad (5.7.11a)$$

$${}^{o2i}\langle\underline{\tilde{E}}_k\rangle = \rho_k {}^{2i}\langle\underline{\tilde{U}}_{kLF}\tilde{E}_{kLF}\rangle \qquad (5.7.11b)$$

$${}^{o2i}\langle\underline{\tilde{E}}_k^T\rangle = 0 \qquad (5.7.11c)$$

$$\gamma_v \alpha_k {}^{ot}\langle\Gamma_k\rangle = \gamma_v \rho_k \left(\frac{\partial \alpha_k}{\partial t} + {}^{3i}\langle\underline{U}_k\rangle_{LF} \cdot \nabla\alpha_k\right) + {}^{o}\langle MTI\rangle_k \qquad (5.7.11d)$$

$${}^{o}\langle MTI\rangle_k = -v^{-1}\rho_k \int_{A_k} \underline{\tilde{U}}_{kLF} \cdot \underline{n}_k dA \qquad (5.7.11e)$$

$$
\begin{aligned}
{}^{o}(TETI)_k = &-\upsilon^{-1}\rho_k \int_{A_k} \tilde{E}_{kLF} \\
&\times ({}^{3i}\langle \underline{U}_k \rangle_{LF} + \underline{\tilde{U}}_{kLF} - \underline{W}_{kLF}) \cdot \underline{n}_k dA \\
&- \upsilon^{-1}\rho_k \int_{A_k} {}^{t}\langle E'_k (\underline{U}'_k - \underline{W}'_k) \rangle \cdot \underline{n}_k dA. \quad (5.7.11f)
\end{aligned}
$$

Substituting Eqs. (5.7.11) to (5.7.11f) into Eq. (5.7.10), we can write the time-volume-averaged total energy equation for ρ_k = constant as follows:

$$
\begin{aligned}
&\gamma_\upsilon \frac{\partial}{\partial t}(\alpha_k {}^{3i}\langle \rho_k \rangle_{LF} {}^{3i}\langle E_k \rangle_{LF}) \\
&\quad + \gamma_A \nabla \cdot [\alpha_k {}^{2i}\langle \rho_k \rangle_{LF} {}^{2i}\langle \underline{U}_k \rangle_{LF} {}^{2i}\langle E_k \rangle_{LF} \\
&\quad + \alpha_k ({}^{o2i}\langle \underline{E}_k^T \rangle + {}^{o2i}\langle \underline{\tilde{E}}_k \rangle)] \\
&= -\gamma_A \nabla \cdot (\alpha_k {}^{2i}\langle \underline{U}_k \rangle_{LF} {}^{2i}\langle P_k \rangle_{LF} + \underline{\Psi}_{Pk}^{2i}) \\
&\quad + \gamma_A \nabla \cdot (\alpha_k {}^{2i}\langle \underline{U}_k \rangle_{LF} \cdot {}^{2i}\langle \underline{\underline{\tau}}_k \rangle_{LF} + \underline{\Psi}_{\tau k}^{2i}) - \gamma_A \nabla \cdot \alpha_k {}^{2i}\langle \underline{J}_{qk} \rangle_{LF} \\
&\quad + \gamma_\upsilon [\alpha_k \rho_k {}^{3i}\langle \underline{U}_k \rangle_{LF} \cdot \underline{f} + \alpha_k {}^{3i}\langle J_{Ek} \rangle_{LF} \\
&\quad + \alpha_k {}^{t3i}\langle \dot{Q}_{kf} \rangle + \alpha_k {}^{t3i}\langle \dot{Q}_{wk} \rangle] \\
&\quad + \gamma_\upsilon {}^{3i}\langle \underline{U}_k \rangle_{LF} {}^{3i}\langle P_k \rangle_{LF} \cdot \nabla \alpha_k - \gamma_\upsilon ({}^{3i}\langle \underline{U}_k \rangle_{LF} \\
&\quad \cdot {}^{3i}\langle \underline{\underline{\tau}}_k \rangle_{LF}) \cdot \nabla \alpha_k + \langle PWI \rangle_k - \langle VWI \rangle_k \\
&\quad + \gamma_\upsilon \alpha_k {}^{ot}\langle \Gamma_k \rangle_{LF} {}^{3i}\langle E_k \rangle_{LF} + {}^{o}\langle TETI \rangle_k. \qquad (5.7.12)
\end{aligned}
$$

5.8 Time-volume-averaged interfacial total energy balance equation (capillary energy ignored)

The local volume-averaged interfacial total energy balance equation is given by Eq. (4.7.3). The time-volume-averaged interfacial total energy balance equation can readily obtained by using Eq. (5.7.9e).

$$\begin{aligned}
&\gamma_v \alpha_k \left({}^{t3i}\langle \dot{Q}_{kf} \rangle + {}^{t3i}\langle \dot{Q}_{wk} \rangle \right) + \gamma_v {}^{3i}\langle \underline{U}_k \rangle_{LF} {}^{3i}\langle P_k \rangle_{LF} \cdot \nabla \alpha_k \\
&- \gamma_v \left({}^{3i}\langle \underline{U}_k \rangle_{LF} \cdot {}^{3i}\langle \underline{\underline{\tau}}_k \rangle_{LF} \right) \cdot \nabla \alpha_k + (PWI)_k - (VWI)_k \\
&+ \gamma_v \alpha_k {}^{t}\langle \Gamma_k \rangle {}^{3i}\langle E_k \rangle_{LF} + (TETI)_k + \gamma_v \alpha_f \left({}^{t3i}\langle \dot{Q}_{fk} \rangle \right. \\
&\left. + {}^{t3i}\langle \dot{Q}_{wf} \rangle \right) + \gamma_v {}^{3i}\langle \underline{U}_f \rangle_{LF} {}^{3i}\langle P_f \rangle_{LF} \cdot \nabla \alpha_f \\
&- \gamma_v \left({}^{3i}\langle \underline{U}_f \rangle_{LF} \cdot {}^{3i}\langle \underline{\underline{\tau}}_f \rangle_{LF} \right) \cdot \nabla \alpha_f + (PWI)_f - (VWI)_f \\
&+ \gamma_v \alpha_f {}^{t}\langle \Gamma_f \rangle {}^{3i}\langle E_f \rangle_{LF} + (TETI)_f = 0. \qquad (5.8.1)
\end{aligned}$$

Here, $(PWI)_k$, $(VWI)_k$, and $(TETI)_k$ are defined in Eqs. (5.7.8b), (5.7.8c), and (5.7.9d). Similarly, for phase f we can write

$$\begin{aligned}
(PWI)_f = &-\upsilon^{-1}\, {}^{3i}\langle P_f \rangle_{LF} \int_{A_f} \underline{\tilde{U}}_{fLF} \cdot \underline{n}_f dA \\
&-\upsilon^{-1}\, {}^{3i}\langle \underline{U}_f \rangle_{LF} \int_{A_f} \tilde{P}_{fLF} \cdot \underline{n}_f dA \\
&-\upsilon^{-1} \int_{A_f} \left(\underline{\tilde{U}}_{fLF} \tilde{P}_{fLF} + {}^{t}\langle \underline{U}'_f P'_f \rangle \right) \cdot \underline{n}_f dA
\end{aligned} \qquad (5.8.2)$$

$$\begin{aligned}
(VWI)_f = &-\upsilon^{-1} \int_{A_f} \left(\underline{\tilde{U}}_{fLF} \cdot {}^{3i}\langle \underline{\underline{\tau}}_f \rangle_{LF} \right) \cdot \underline{n}_f dA \\
&-\upsilon^{-1} \int_{A_f} \left({}^{3i}\langle \underline{U}_f \rangle_{LF} \cdot \underline{\underline{\tilde{\tau}}}_{fLF} \right) \cdot \underline{n}_f dA \\
&-\upsilon^{-1} \int_{A_f} \left(\underline{\tilde{U}}_{fLF} \cdot \underline{\underline{\tilde{\tau}}}_{fLF} + {}^{t}\langle \underline{U}'_f \cdot \underline{\underline{\tau}}'_f \rangle \right) \cdot \underline{n}_f dA
\end{aligned} \qquad (5.8.3)$$

$$
\begin{aligned}
&(TETI)_f \\
&= -\upsilon^{-1}\int_{A_f} \tilde{E}_{fLF}({}^{3i}\langle\rho_f\rangle_{LF} + \tilde{\rho}_{fLF}) \\
&\quad \times ({}^{3i}\langle\underline{U}_f\rangle_{LF} + \underline{\tilde{U}}_{fLF} - \underline{W}_{fLF})\cdot\underline{n}_f dA \\
&\quad - \upsilon^{-1}\int_{A_f} ({}^{3i}\langle\rho_f\rangle_{LF} + \tilde{\rho}_{fLF})\,{}^{t}\langle E'_f(\underline{U}'_f - \underline{W}'_f)\rangle\cdot\underline{n}_f dA \\
&\quad - \upsilon^{-1}\int_{A_f} \tilde{E}_{fLF}\,{}^{t}\langle\rho'_f(\underline{U}'_f - \underline{W}'_f)\rangle\cdot\underline{n}_f dA \\
&\quad - \upsilon^{-1}\int_{A_f} ({}^{3i}\langle\underline{U}_f\rangle_{LF} + \underline{\tilde{U}}_{fLF} - \underline{W}_{fLF})\,{}^{t}\langle\rho'_f E'_f\rangle\cdot\underline{n}_f dA \\
&\quad - \upsilon^{-1}\int_{A_f} {}^{t}\langle\rho'_f E'_f(\underline{U}'_f - \underline{W}'_f)\rangle\cdot\underline{n}_f dA. \qquad (5.8.4)
\end{aligned}
$$

5.9 Time-volume-averaged internal energy conservation equation

The local volume-averaged internal energy conservation equation was given by Eq. (4.6.4):

$$
\begin{aligned}
&\gamma_\upsilon\frac{\partial}{\partial t}\alpha_k\,{}^{3i}\langle\rho_k u_k\rangle + \gamma_A\nabla\cdot\alpha_k\,{}^{2i}\langle\rho_k\underline{U}_k u_k\rangle \\
&\quad = -\gamma_\upsilon\alpha_k\,{}^{3i}\langle P_k\nabla\cdot\underline{U}_k\rangle - \gamma_A\nabla\cdot\alpha_k\,{}^{2i}\langle\underline{J}_{qk}\rangle \\
&\qquad + \gamma_\upsilon\alpha_k({}^{3i}\langle\Phi_k\rangle + {}^{3i}\langle J_{Ek}\rangle + {}^{3i}\langle\dot{Q}_{kf}\rangle + {}^{3i}\langle\dot{Q}_{wk}\rangle) \\
&\qquad - \upsilon^{-1}\int_{A_k}\rho_k u_k(\underline{U}_k - \underline{W}_k)\cdot\underline{n}_k dA, \qquad (4.6.4)
\end{aligned}
$$

in which

$$
\begin{aligned}
&{}^{t}\left\langle -\upsilon^{-1}\int_{A_{kf}}\underline{J}_{qk}\cdot\underline{n}_k dA\right\rangle = -\upsilon^{-1}\int_{A_{kf}}{}^{t}\langle\underline{J}_{qk}\rangle\cdot\underline{n}_k dA \\
&= \gamma_\upsilon\,{}^{3i}\langle\underline{J}_{qk}\rangle_{LF}\cdot\nabla\alpha_k - \upsilon^{-1}\int_{A_{kf}}\underline{\tilde{J}}_{qkLF}\cdot\underline{n}_k dA \\
&= \gamma_\upsilon\,{}^{3i}\langle\underline{J}_{qk}\rangle_{LF}\cdot\nabla\alpha_k + (HTI)_k = \gamma_\upsilon\alpha_k\,{}^{t3i}\langle\dot{Q}_{kf}\rangle \qquad (5.7.7\text{c,d})
\end{aligned}
$$

$$\gamma_v \alpha_k {}^{t3i}\langle \dot{Q}_{wk} \rangle = -v^{-1} \int_{A_{wk}} \underline{J}_{qk} \cdot \underline{n}_k dA. \quad (5.7.7e)$$

Previously, we also give Φ_k, the dissipation function,

$$\Phi_k = \underline{\underline{\tau}}_k : \nabla, \underline{U}_k, \quad (4.6.4a)$$

in which the double dot denotes the scalar product of two second-order tensors, the comma denotes the dyadic operation, and Φ_k gives the dissipation rate per unit volume of phase k relating from the irreversible conversion of mechanical work into thermal energy. Time averaging of the local volume-averaged internal energy conservation equation requires consideration of the following:

The first term on the LHS of Eq. (4.6.4) is

$$\gamma_v \frac{\partial}{\partial t} {}^t\langle \alpha_k {}^{3i}\langle \rho_k u_k \rangle\rangle \quad (5.9.1)$$

$${}^t\langle \alpha_k {}^{3i}\langle \rho_k u_k \rangle\rangle = \alpha_k {}^{3i}\langle \rho_k \rangle_{LF} {}^{3i}\langle u_k \rangle_{LF} + \phi^{3i}_{uk}, \quad (5.9.1a)$$

in which ϕ^{3i}_{uk} is a scalar internal energy function defined by

$$\phi^{3i}_{uk} = \alpha_k {}^{3i}\langle \tilde{\rho}_{kLF} \tilde{u}_{kLF} \rangle + \alpha_k {}^{t3i}\langle \rho'_k u'_k \rangle. \quad (5.9.1b)$$

The first term on the LHS of Eq. (4.6.4) can be written as

$$\gamma_v \frac{\partial}{\partial t} {}^t\langle \alpha_k {}^{3i}\langle \rho_k u_k \rangle\rangle = \gamma_v \frac{\partial}{\partial t} \left(\alpha_k {}^{3i}\langle \rho_k \rangle_{LF} {}^{3i}\langle u_k \rangle_{LF} + \phi^{3i}_{uk} \right). \quad (5.9.1c)$$

The second term on the LHS of Eq. (4.6.4) is

$$\gamma_A \nabla \cdot {}^t\langle \alpha_k {}^{2i}\langle \rho_k \underline{U}_k u_k \rangle\rangle \quad (5.9.2)$$

$$
\begin{aligned}
{}^{t}\langle \alpha_k\, {}^{2i}\langle \rho_k \underline{U}_k u_k\rangle\rangle &= {}^{t}\langle \alpha_k\, {}^{2i}\langle ({}^{2i}\langle \rho_k\rangle_{LF} + \tilde{\rho}_{kLF} + \rho'_k)({}^{2i}\langle \underline{U}_k\rangle_{LF} \\
&\quad + \tilde{\underline{U}}_{kLF} + \underline{U}'_k)({}^{2i}\langle u_k\rangle_{LF} + \tilde{u}_{kLF} + u'_k)\rangle\rangle \\
&= \alpha_k\, {}^{2i}\langle \rho_k\rangle_{LF}\, {}^{2i}\langle \underline{U}_k\rangle_{LF}\, {}^{2i}\langle u_k\rangle_{LF} \\
&\quad + {}^{2i}\langle \underline{U}_k\rangle_{LF}\, \phi^{2i}_{uk} + {}^{2i}\langle u_k\rangle_{LF}\, \underline{\psi}^{2i}_{mk} \\
&\quad + \alpha_k \left({}^{2i}\langle \underline{u}^T_k\rangle + {}^{2i}\langle \tilde{\underline{u}}_k\rangle + {}^{2i}\langle \tilde{\underline{u}}^T_k\rangle\right), \qquad (5.9.2a)
\end{aligned}
$$

$$\text{where } \phi^{2i}_{uk} = \alpha_k\, {}^{2i}\langle \tilde{\rho}_{kLF} \tilde{u}_{kLF}\rangle + \alpha_k\, {}^{t2i}\langle \rho'_k u'_k\rangle. \qquad (5.9.2b)$$

$\underline{\psi}^{2i}_{mk}$ is defined as in Eq. (5.3.2h).

(a) The volume-averaged turbulent internal energy flux ${}^{2i}\langle \underline{u}^T_k\rangle$ is defined by

$$ {}^{2i}\langle \rho_k\rangle_{LF}\, {}^{t2i}\langle \underline{U}'_k u'_k\rangle + {}^{t2i}\langle \tilde{\rho}_{kLF} \underline{U}'_k u'_k\rangle. \qquad (5.9.2c)$$

Here, ${}^{3i}\langle \underline{u}^T_k\rangle$ can be expressed in terms of eddy diffusivity for internal energy transfer D^T_{uk} according to

$$ {}^{2i}\langle \underline{u}^T_k\rangle = -\, {}^{2i}\langle \rho_k\rangle_{LF} D^T_{uk} \nabla\, {}^{2i}\langle u_k\rangle_{LF} \qquad (5.9.2d)$$

because $\nabla\, {}^{2i}\langle u_k\rangle_{LF} = c_{vk} \nabla\, {}^{2i}\langle T_k\rangle_{LF}$, where c_{vk} is the specific heat at constant volume (assumed constant in the present analysis) and T is the temperature, respectively. Equation (5.9.2d) can be written as

$$ {}^{2i}\langle \underline{u}^T_k\rangle = -\kappa^T_k \nabla\, {}^{2i}\langle T_k\rangle_{LF}, \qquad (5.9.2e)$$

where κ^T_k is the turbulent conductivity related to D^T_{uk} according to

$$\kappa^T_k = {}^{2i}\langle \rho_k\rangle_{LF} c_{vk} D^T_{uk}. \qquad (5.9.2f)$$

(b) The volume-averaged dispersive internal energy flux ${}^{2i}\langle \underline{\tilde{u}}_k \rangle$ is defined by

$$^{2i}\langle \rho_k \rangle_{LF}\, {}^{2i}\langle \underline{\tilde{U}}_{kLF} \tilde{u}_{kLF} \rangle + {}^{2i}\langle \tilde{\rho}_{kLF} \underline{\tilde{U}}_{kLF} \tilde{u}_{kLF} \rangle. \qquad (5.9.2g)$$

(c) The volume-averaged turbulent, dispersive internal energy flux ${}^{2i}\langle \underline{\tilde{u}}_k^T \rangle$ is defined by

$$^{t2i}\langle \underline{\tilde{U}}_{kLF} \rho'_k u'_k \rangle + {}^{t2i}\langle \tilde{u}_{kLF} \rho'_k \underline{U}'_k \rangle. \qquad (5.9.2h)$$

(d) $\rho'_k \underline{U}'_k u'_k$ is a triple time correlation that is assumed to be small and can be neglected. (5.9.2i)

The second term on the LHS of Eq. (4.6.4) thus can be written as

$$\begin{aligned} &\gamma_A \nabla \cdot {}^t\langle \alpha_k\, {}^{2i}\langle \rho_k \underline{U}_k u_k \rangle\rangle \\ &= \gamma_A \nabla \cdot \big[\alpha_k\, {}^{2i}\langle \rho_k \rangle_{LF}\, {}^{2i}\langle \underline{U}_k \rangle_{LF}\, {}^{2i}\langle u_k \rangle_{LF} + {}^{2i}\langle \underline{U}_k \rangle_{LF} \phi_{uk}^{2i} \\ &\quad + {}^{2i}\langle u_k \rangle_{LF} \underline{\psi}_{mk}^{2i} + \alpha_k \big({}^{2i}\langle \underline{u}_k^T \rangle + {}^{2i}\langle \underline{\tilde{u}}_k \rangle + {}^{2i}\langle \underline{\tilde{u}}_k^T \rangle\big)\big]. \quad (5.9.2j) \end{aligned}$$

The first term on the RHS of Eq. (4.6.4) is

$$-\gamma_v\, {}^t\langle \alpha_k\, {}^{3i}\langle P_k \nabla \cdot \underline{U}_k \rangle\rangle \qquad (5.9.3)$$

$$\begin{aligned} &{}^t\langle \alpha_k\, {}^{3i}\langle P_k \nabla \cdot \underline{U}_k \rangle\rangle \\ &= {}^t\langle \alpha_k\, {}^{3i}\langle ({}^{3i}\langle P_k \rangle_{LF} + \tilde{P}_{kLF} + P'_k) \\ &\quad \times (\nabla \cdot {}^{3i}\langle {}^{3i}\langle \underline{U}_k \rangle_{LF} + \underline{\tilde{U}}_{kLF} + \underline{U}'_k \rangle)\rangle\rangle \\ &= \alpha_k ({}^{3i}\langle P_k \rangle_{LF} \nabla \cdot {}^{3i}\langle \underline{U}_k \rangle_{LF} + {}^{3i}\langle \tilde{P}_{kLF} \nabla \cdot \underline{\tilde{U}}_{kLF} \rangle \\ &\quad + {}^{t3i}\langle P'_k \nabla \cdot \underline{U}'_k \rangle) \\ &\quad + {}^t\langle \alpha_k\, {}^{3i}\langle \tilde{P}_{kLF} \nabla \cdot \underline{U}'_k \rangle\rangle + {}^t\langle \alpha_k\, {}^{3i}\langle P'_k \nabla \cdot \underline{\tilde{U}}_{kLF} \rangle\rangle \\ &\quad + {}^{3i}\langle \alpha_k\, {}^{3i}\langle P_k \rangle_{LF} \nabla \cdot \underline{\tilde{U}}_{kLF} \rangle + {}^{3i}\langle \alpha_k \tilde{P}_{kLF} \nabla \cdot {}^{3i}\langle \underline{U}_k \rangle_{LF} \rangle \\ &\quad + {}^t\langle \alpha_k\, {}^{3i}\langle P'_k \nabla \cdot {}^{3i}\langle \underline{U}_k \rangle_{LF} \rangle\rangle + {}^t\langle \alpha_k\, {}^{3i}\langle {}^{3i}\langle P_k \rangle_{LF} \nabla \cdot \underline{U}'_k \rangle\rangle. \end{aligned}$$

(5.9.3a)

After time averaging of Eq. (5.9.3a), Eq. (5.9.3) becomes

$$-\gamma_v\,{}^t\langle\alpha_k\,{}^{3i}\langle P_k\nabla\cdot\underline{U}_k\rangle\rangle = -\gamma_v[\alpha_k({}^{3i}\langle P_k\rangle_{LF}\nabla\cdot{}^{3i}\langle\underline{U}_k\rangle_{LF} + \phi_{Pk}^{3i})] + (PWI)_k^{(u)}, \qquad (5.9.3b)$$

where ϕ_{Pk}^{3i} is a scalar pressure work function defined as

$$\phi_{Pk}^{3i} = \alpha_k\,{}^{3i}\langle\tilde{P}_{kLF}\nabla\cdot\underline{\tilde{U}}_{kLF}\rangle + \alpha_k\,{}^{t3i}\langle P_k'\nabla\cdot\underline{U}_k'\rangle. \quad (5.9.3c)$$

Here, $(PWI)_k^{(u)}$ denotes a portion of the interfacial pressure work integral that is shown in Eq. (5.7.8b) and defined here as

$$(PWI)_k^{(u)} = -\upsilon^{-1}\,{}^{3i}\langle P_k\rangle_{LF}\int_{A_k}\underline{\tilde{U}}_{kLF}\cdot\underline{n}_k dA. \quad (5.9.3d)$$

The superscript (u) is a reminder that is associated with internal energy.

The first term on the RHS of Eq. (4.6.4) can be written as follows:

$$-\gamma_v\,{}^t\langle\alpha_k\,{}^{3i}\langle P_k\nabla\cdot\underline{U}_k\rangle\rangle = -\gamma_v\,(\alpha_k\,{}^{3i}\langle P_k\rangle_{LF}\nabla\cdot{}^{3i}\langle\underline{U}_k\rangle_{LF} + \phi_{Pk}^{3i}) + (PWI)_k^{(u)}. \qquad (5.9.3e)$$

The second term on the RHS of the local volume-averaged internal energy conservation equation [Eq. (4.6.4)] is identical to the third term on the RHS of the local volume-averaged total energy conservation equation [Eq. (4.6.3)]. Equations (5.7.6c) and (5.7.6d), give

$$-\gamma_A\nabla\cdot\alpha_k\,{}^{2i}\langle\underline{J}_{qk}\rangle = \gamma_A\nabla\cdot\alpha_k\frac{\kappa_k}{c_{\upsilon k}}\nabla\,{}^{2i}\langle u_k\rangle_{LF} = -\gamma_A\nabla\cdot\alpha_k\,{}^{2i}\langle\underline{J}_{qk}\rangle_{LF}. \qquad (5.9.4)$$

The third term on the RHS of Eq. (4.6.4) is almost identical to the fourth term on the RHS of Eq. (4.6.3), except that the dissipation function $\Phi_k = \underline{\underline{\tau}}_k : \nabla, \underline{U}_k$ in Eq. (4.6.4b) replaces ${}^{3i}\langle \rho_k \underline{U}_k \rangle \cdot \underline{f}$ in Eq. (4.6.3),

$$\begin{aligned}
\gamma_v {}^t\langle \alpha_k {}^{3i}\langle \Phi_k \rangle\rangle &= \gamma_v {}^t\langle \alpha_k {}^{3i}\langle \underline{\underline{\tau}}_k : \nabla, \underline{U}_k \rangle\rangle \\
&= \gamma_v {}^t\langle \alpha_k [({}^{3i}\langle \underline{\underline{\tau}}_k \rangle_{LF} + \underline{\underline{\tilde{\tau}}}_{kLF} + \underline{\underline{\tau}}'_k) : (\nabla, {}^{3i}\langle \underline{U}_k \rangle_{LF} \\
&\quad + \nabla, \underline{\tilde{U}}_{kLF} + \nabla, \underline{U}'_k)]\rangle \\
&= \gamma_v (\alpha_k {}^{3i}\langle \underline{\underline{\tau}}_k \rangle_{LF} : \nabla, {}^{3i}\langle \underline{U}_k \rangle_{LF} + \phi^{3i}_{\tau k}) \\
&\quad + (VDI)_k,
\end{aligned} \tag{5.9.5}$$

where $\phi^{3i}_{\tau k}$ is a scalar viscous dissipation function defined as

$$\phi^{3i}_{\tau k} = \alpha_k {}^{3i}\langle \underline{\underline{\tilde{\tau}}}_{kLF} : \nabla, \underline{\tilde{U}}_{kLF} \rangle + \alpha_k {}^{t3i}\langle \underline{\underline{\tau}}'_k : \nabla, \underline{U}'_k \rangle \tag{5.9.5a}$$

and $(VDI)_k$ is an interfacial viscous dissipation integral defined as

$$(VDI)_k = v^{-1}\, {}^{3i}\langle \underline{\underline{\tau}}_k \rangle_{LF} : \int_{A_k} \underline{\tilde{U}}_{kLF}, \underline{n}_k dA. \tag{5.9.5b}$$

The similarity between Eqs. (5.9.3b) and (5.9.5) is understandable if one recalls that $\Phi_k = \underline{\underline{\tau}}_k : \nabla, \underline{U}_k$. Physically, Eq. (5.9.3b) gives the reversible conversion of mechanical work into thermal energy, and may be either positive or negative. Equation (5.9.5) gives the irreversible conversion of mechanical work into thermal energy and is always positive.

The remaining parts of the third term on the RHS of Eq. (4.6.4) were given by Eq. (5.7.7b) for $\gamma_v \alpha_k {}^{3i}\langle J_{Ek} \rangle_{LF}$, Eq. (5.7.7c) for $\gamma_v \alpha_k {}^{t3i}\langle \dot{Q}_{kf} \rangle$, and Eq. (5.7.7e) for $\gamma_v \alpha_k {}^{t3i}\langle \dot{Q}_{wk} \rangle$.

The third term on the RHS of Eq. (4.6.4) can be written as

$$\gamma_v \alpha_k ({}^{t3i}\langle \Phi_k \rangle + {}^{3i}\langle J_{Ek} \rangle_{LF} + {}^{t3i}\langle \dot{Q}_{kf} \rangle + {}^{t3i}\langle \dot{Q}_{wk} \rangle). \quad (5.9.5c)$$

The fourth term on the RHS of Eq. (4.6.4) is

$$\begin{aligned} &{}^{t}\left\langle -v^{-1} \int_{A_k} \rho_k u_k (\underline{U}_k - \underline{W}_k) \cdot \underline{n}_k dA \right\rangle \\ &\quad = -v^{-1} \int_{A_k} {}^{t}\langle \rho_k u_k (\underline{U}_k - \underline{W}_k) \cdot \underline{n}_k dA \rangle. \end{aligned} \quad (5.9.6)$$

$$\begin{aligned} &-v^{-1} \int_{A_k} {}^{t}\langle \rho_k u_k (\underline{U}_k - \underline{W}_k) \cdot \underline{n}_k dA \rangle \\ &\quad = -v^{-1} \int_{A_k} {}^{t}\langle ({}^{3i}\langle \rho_k \rangle_{LF} + \tilde{\rho}_{kLF} + \rho_k')({}^{3i}\langle u_k \rangle_{LF} \\ &\quad + \tilde{u}_{kLF} + u_k')(({}^{3i}\langle \underline{U}_k \rangle_{LF} + \underline{\tilde{U}}_{kLF} + \underline{U}_k') \\ &\quad - (\underline{W}_{kLF} + \underline{W}_k')) \cdot \underline{n}_k dA \rangle. \end{aligned} \quad (5.9.6a)$$

Because low- and high-frequency parts of Eq. (5.9.6) can be separated, after time averaging the equation becomes

$$\begin{aligned} &-v^{-1} \int_{A_k} {}^{t}\langle \rho_k u_k (\underline{U}_k - \underline{W}_k) \rangle \cdot \underline{n}_k dA \\ &= \gamma_v {}^{3i}\langle \rho_k \rangle_{LF} {}^{3i}\langle u_k \rangle_{LF} \left(\frac{\partial \alpha_k}{\partial t} + {}^{3i}\langle \underline{U}_k \rangle_{LF} \cdot \nabla \alpha_k \right) \\ &\quad - v^{-1\,3i}\langle \rho_k \rangle_{LF} {}^{3i}\langle u_k \rangle_{LF} \int_{A_k} \underline{\tilde{U}}_{kLF} \cdot \underline{n}_k dA \\ &\quad - v^{-1\,3i}\langle u_k \rangle_{LF} \int_{A_k} \tilde{\rho}_{kLF} ({}^{3i}\langle \underline{U}_k \rangle_{FL} + \underline{\tilde{U}}_{kLF} - \underline{W}_{kLF}) \cdot \underline{n}_k dA \\ &\quad - v^{-1} \int_{A_k} ({}^{3i}\langle \rho_k \rangle_{LF} + \tilde{\rho}_{kLF}) \tilde{u}_{kLF} ({}^{3i}\langle \underline{U}_k \rangle_{LF} \\ &\quad + \underline{\tilde{U}}_{kLF} - \underline{W}_{kLF}) \cdot \underline{n}_k dA \end{aligned}$$

$$- \upsilon^{-1} \int_{A_k} ({}^{3i}\langle \rho_k \rangle_{LF} + \tilde{\rho}_{kLF})^t \langle u'_k (\underline{U}'_k - \underline{W}'_k) \rangle \cdot \underline{n}_k dA$$
$$- \upsilon^{-1} \int_{A_k} ({}^{3i}\langle u_k \rangle_{LF} + \tilde{u}_{kLF})^t \langle \rho'_k (\underline{U}'_k - \underline{W}'_k) \rangle \cdot \underline{n}_k dA$$
$$- \upsilon^{-1t} \left\langle \int_{A_k} ({}^{3i}\langle \underline{U}_k \rangle_{LF} + \tilde{\underline{U}}_{kLF} - \underline{W}_{kLF})^t \langle \rho'_k u'_k \rangle \cdot \underline{n}_k dA \right\rangle$$
$$- \upsilon^{-1} \int_{A_k} {}^t\langle \rho'_k u'_k (\underline{U}'_k - \underline{W}'_k) \rangle \cdot \underline{n}_k dA. \qquad (5.9.6b)$$

The fourth terms on the RHS of Eq. (4.6.4) can be written by multiplying Eq. (5.3.3c) by ${}^{3i}\langle u_k \rangle_{LF}$. Introducing the results into Eq. (5.9.6b) then leads to

$$-\upsilon^{-1} \int_{A_k} {}^t\langle \rho_k u_k (\underline{U}_k - \underline{W}_k) \rangle \cdot \underline{n}_k dA$$
$$= \gamma_\upsilon \alpha_k {}^t\langle \Gamma_k \rangle {}^{3i}\langle u_k \rangle_{LF} + (IETI)_k. \qquad (5.9.6c)$$

The interfacial internal energy transfer integral is

$$(IETI)_k = -\upsilon^{-1} \int_{A_k} \tilde{u}_{kLF} ({}^{3i}\langle \rho_k \rangle_{LF} + \tilde{\rho}_{kLF}) ({}^{3i}\langle \underline{U}_k \rangle_{LF}$$
$$+ \tilde{\underline{U}}_{kLF} - \underline{W}_{kLF}) \cdot \underline{n}_k dA$$
$$-\upsilon^{-1} \int_{A_k} ({}^{3i}\langle \rho_k \rangle_{LF} + \tilde{\rho}_{kLF})^t \langle u'_k (\underline{U}'_k - \underline{W}'_k) \rangle \cdot \underline{n}_k dA$$
$$-\upsilon^{-1} \int_{A_k} \tilde{u}_{kLF} {}^t\langle \rho'_k (\underline{U}'_k - \underline{W}'_k) \rangle \cdot \underline{n}_k dA$$
$$-\upsilon^{-1} \int_{A_k} ({}^{3i}\langle \underline{U}_k \rangle_{LF} + \tilde{\underline{U}}_{kLF} - \underline{W}_{kLF})^t \langle \rho'_k u'_k \rangle \cdot \underline{n}_k dA$$
$$-\upsilon^{-1} \int_{A_k} {}^t\langle \rho'_k u'_k (\underline{U}'_k - \underline{W}'_k) \rangle \cdot \underline{n}_k dA. \qquad (5.9.6d)$$

An examination of the preceding results shows that the time-averaged total interfacial internal energy source of phase k per unit volume in v, $\gamma_v \alpha_k {}^t\langle\Theta_k\rangle$ consists of (1) a portion of the interfacial pressure work $(PWI)_k^{(u)}$ defined by Eq. (5.9.3d), (2) the interfacial dissipation integral $(VDI)_k$ defined by Eq. (5.9.5b), and (3) the interfacial heat transfer $\gamma_v \alpha_k ({}^{t3i}\langle\dot{Q}_{kf}\rangle + {}^{t3i}\langle\dot{Q}_{wk}\rangle)$ defined by Eqs. (4.3.8), and (4.3.9), respectively, and (4) the interfacial internal energy transfer given by Eq. (5.9.6c) related to the mass generation. Hence,

$$\gamma_v \alpha_k {}^t\langle\Theta_k\rangle = (PWI)_k^{(u)} + (VDI)_k + \gamma_v \alpha_k ({}^{t3i}\langle\dot{Q}_{kf}\rangle + {}^{t3i}\langle\dot{Q}_{wk}\rangle) + \gamma_v \alpha_k {}^t\langle\Gamma_k\rangle {}^{3i}\langle u_k\rangle_{LF} + (IETI)_k. \tag{5.9.6e}$$

By using the results given in Eqs. (5.9.1c), (5.9.2j), (5.9.3e), (5.9.4), (5.9.5), (5.7.7b), (5.7.7c), (5.7.7e), and (5.9.6c), we obtain the time volume-averaged internal energy conservation equation

$$\begin{aligned}
&\gamma_v \frac{\partial}{\partial t}(\alpha_k {}^{3i}\langle\rho_k\rangle_{LF} {}^{3i}\langle u_k\rangle_{LF} + \phi^{3i}_{uk}) \\
&\quad + \gamma_A \nabla \cdot [\alpha_k {}^{2i}\langle\rho_k\rangle_{LF} {}^{2i}\langle\underline{U}_k\rangle_{LF} {}^{2i}\langle u_k\rangle_{LF} + \phi^{2i}_{uk} {}^{2i}\langle\underline{U}_k\rangle_{LF} \\
&\quad + \underline{\psi}^{2i}_{mk} {}^{2i}\langle u_k\rangle_{LF} + \alpha_k({}^{2i}\langle\underline{u}_k^T\rangle + {}^{2i}\langle\tilde{\underline{u}}_k\rangle + {}^{2i}\langle\tilde{\underline{u}}_k^T\rangle)] \\
&= -\gamma_v(\alpha_k {}^{3i}\langle P_k\rangle_{LF} \nabla \cdot {}^{3i}\langle\underline{U}_k\rangle_{LF} + \phi^{3i}_{Pk}) \\
&\quad + (PWI)_k^{(u)} - \gamma_A \nabla \cdot \alpha_k {}^{2i}\langle\underline{J}_{qk}\rangle_{LF} \\
&\quad + \gamma_v(\alpha_k {}^{3i}\langle\underline{\underline{\tau}}_k\rangle_{LF} : \nabla, {}^{3i}\langle\underline{U}_k\rangle_{LF} + \phi^{3i}_{\tau k}) + (VDI)_k \\
&\quad + \gamma_v \alpha_k({}^{3i}\langle J_{Ek}\rangle_{LF} + {}^{t3i}\langle\dot{Q}_{kf}\rangle + {}^{t3i}\langle\dot{Q}_{wk}\rangle) \\
&\quad + \gamma_v \alpha_k {}^t\langle\Gamma_k\rangle {}^{3i}\langle u_k\rangle_{LF} + (IETI)_k.
\end{aligned} \tag{5.9.7}$$

When ρ_k = constant, ${}^{3i}\langle\rho_k\rangle_{LF} = \rho_k$, $\tilde{\rho}_{kLF} = \rho'_k = 0$. Therefore,

$$\phi^{3i}_{uk} = 0,\ \phi^{2i}_{uk} = 0,\ \underline{\psi}^{3i}_{mk} = 0 \text{ and } \underline{\psi}^{2i}_{mk} = 0 \quad (5.9.8)$$

$$^{o2i}\langle \underline{u}^T_k\rangle = \rho_k\, {}^{t2i}\langle \underline{U}'_k u'_k\rangle \quad (5.9.8a)$$

$$^{o2i}\langle \underline{\tilde{u}}_k\rangle = \rho_k\, {}^{2i}\langle \underline{\tilde{U}}_{kLF}\tilde{u}_{kLF}\rangle \quad (5.9.8b)$$

$$^{o2i}\langle \underline{\tilde{u}}^T_k\rangle = 0 \quad (5.9.8c)$$

$$\gamma_v\alpha_k\, {}^{ot}\langle\Gamma_k\rangle = \gamma_v\rho_k\left(\frac{\partial\alpha_k}{\partial t} + {}^{3i}\langle\underline{U}_k\rangle_{LF}\cdot\nabla\alpha_k\right) + {}^{o}\langle MTI\rangle_k \quad (5.9.8d)$$

$$^{o}\langle MTI\rangle_k = -v^{-1}\rho_k\int_{A_k}\underline{\tilde{U}}_{kLF}\cdot\underline{n}_k dA \quad (5.9.8e)$$

$$^{o}(IETI)_k = -v^{-1}\rho_k\int_{A_k}\tilde{u}_k({}^{3i}\langle\underline{U}_k\rangle_{LF} + \underline{\tilde{U}}_{kLF} - \underline{W}_{kLF})\cdot\underline{n}_k dA - v^{-1}\rho_k\int_{A_k} u'_k(\underline{U}'_k - \underline{W}'_k)\cdot\underline{n}_k dA. \quad (5.9.8f)$$

Substituting Eqs. (5.9.8–5.9.8f) into Eq. (5.9.7), we can write the time-volume-averaged internal energy equation for ρ_k = constant as follows.

$$\begin{aligned}
&\gamma_v\frac{\partial}{\partial t}(\alpha_k\rho_k\, {}^{3i}\langle u_k\rangle_{LF}) + \gamma_A\nabla\cdot\left[\alpha_k\rho_k\, {}^{3i}\langle\underline{U}_k\rangle_{LF}\, {}^{3i}\langle u_k\rangle_{LF}\right.\\
&\quad\left. + \alpha_k\left({}^{o2i}\langle\underline{u}^T_k\rangle + {}^{o2i}\langle\underline{\tilde{u}}_k\rangle\right)\right]\\
&= -\gamma_v\left[\alpha_k\, {}^{3i}\langle P_k\rangle_{LF}\nabla\cdot{}^{3i}\langle\underline{U}_k\rangle_{LF} + \phi^{3i}_{Pk}\right]\\
&\quad + (PWI)^{(u)}_k - \gamma_A\nabla\cdot\alpha_k\, {}^{2i}\langle\underline{J}_{qk}\rangle_{LF}\\
&\quad + \gamma_v\left[\alpha_k\, {}^{3i}\langle\underline{\underline{\tau}}_k\rangle_{LF} : \nabla,\ {}^{3i}\langle\underline{U}_k\rangle_{LF} + \phi^{3i}_{\tau k}\right]\\
&\quad + (VDI)_k + \gamma_v\alpha_k\left[{}^{3i}\langle J_{Ek}\rangle_{LF} + {}^{t3i}\langle\dot{Q}_{kf}\rangle + {}^{t3i}\langle\dot{Q}_{wk}\rangle\right]\\
&\quad + \gamma_v\alpha_k\, {}^{ot}\langle\Gamma_k\rangle\, {}^{3i}\langle u_k\rangle_{LF} + {}^{o}(IETI)_k.
\end{aligned} \quad (5.9.9)$$

5.10 Time-volume-averaged interfacial internal energy balance equation

The local volume-averaged interfacial internal energy balance equation is given by Eq. (4.7.4). The time-volume-averaged interfacial internal energy balance equation can be readily obtained by using Eq. (5.9.6e):

$$\begin{aligned}(PWI)_k^{(u)} &+ (VDI)_k + \gamma_v \alpha_k ({}^{t3i}\langle \dot{Q}_{kf}\rangle + {}^{t3i}\langle \dot{Q}_{wk}\rangle)\\ &+ \gamma_v \alpha_k {}^t\langle \Gamma_k \rangle^{3i} \langle u_k \rangle_{LF} + (IETI)_k\\ &+ (PWI)_f^{(u)} + (VDI)_f + \gamma_v \alpha_f ({}^{t3i}\langle \dot{Q}_{fk}\rangle + {}^{t3i}\langle \dot{Q}_{wf}\rangle)\\ &+ \gamma_v \alpha_f {}^t\langle \Gamma_f \rangle^{3i} \langle u_f \rangle_{LF} + (IETI)_f = 0. \qquad (5.10.1)\end{aligned}$$

The definitions of $(PWI)_k^{(u)}$, $(VDI)_k$, and $(IETI)_k$ are shown in Eqs. (5.9.3d), (5.9.5b), and (5.9.6d). Similarly for phase f, we can write

$$(PWI)_f^{(u)} = -\upsilon^{-1\,3i}\langle P_f \rangle_{LF} \int_{A_f} \underline{\tilde{U}}_{fLF} \cdot \underline{n}_f dA \qquad (5.10.2)$$

$$(VDI)_f = \upsilon^{-1\,3i}\langle \underline{\underline{\tau}}_f \rangle_{LF} : \int_{A_f} \underline{\tilde{U}}_{fLF}, \underline{n}_f dA \qquad (5.10.3)$$

$$\begin{aligned}(IETI)_f = -\upsilon^{-1} &\int_{A_f} \tilde{u}_{fLF}({}^{3i}\langle \rho_f \rangle_{LF} + \tilde{\rho}_{fLF})({}^{3i}\langle \underline{U}_f \rangle_{LF}\\ &+ \underline{\tilde{U}}_{fLF} - \underline{W}_{fLF}) \cdot \underline{n}_f dA\\ &- \upsilon^{-1} \int_{A_f} ({}^{3i}\langle \rho_f \rangle_{LF} + \tilde{\rho}_{fLF}) {}^t\langle u'_f(\underline{U}'_f - \underline{W}'_f)\rangle \cdot \underline{n}_f dA\\ &- \upsilon^{-1} \int_{A_f} \tilde{u}_{fLF} {}^t\langle \rho'_f(\underline{U}'_f - \underline{W}'_f)\rangle \cdot \underline{n}_f dA\\ &- \upsilon^{-1} \int_{A_f} ({}^{3i}\langle \underline{U}_f \rangle_{LF} + \underline{\tilde{U}}_{fLF} - \underline{W}_{fLF}) {}^t\langle \rho'_f u'_f \rangle \cdot \underline{n}_f dA\\ &- \upsilon^{-1} \int_{A_f} {}^t\langle \rho'_f u'_f(\underline{U}'_f - \underline{W}'_f)\rangle \cdot \underline{n}_f dA. \qquad (5.10.4)\end{aligned}$$

It is to be noted that both internal energy dissipation and reversible work are shown in the interfacial internal energy balance equation.

5.11 Time-volume-averaged enthalpy conservation equation

The local volume-averaged enthalpy conservation equation is given by Eq. (4.6.5):

$$\gamma_v \frac{\partial}{\partial t} \alpha_k {}^{3i}\langle \rho_k h_k \rangle + \gamma_A \nabla \cdot \alpha_k {}^{2i}\langle \rho_k \underline{U}_k h_k \rangle$$
$$= \gamma_v \frac{\partial}{\partial t} \alpha_k {}^{3i}\langle P_k \rangle_{LF} + \gamma_A \nabla \cdot \alpha_k {}^{2i}\langle P_k \underline{U}_k \rangle$$
$$- \gamma_v \alpha_k {}^{3i}\langle P_k \nabla \cdot \underline{U}_k \rangle - \gamma_A \nabla \cdot \alpha_k {}^{2i}\langle \underline{J}_{qk} \rangle$$
$$+ \gamma_v \alpha_k ({}^{3i}\langle \Phi_k \rangle + {}^{3i}\langle J_{Ek} \rangle + {}^{3i}\langle \dot{Q}_{kf} \rangle + {}^{3i}\langle \dot{Q}_{wk} \rangle)$$
$$+ v^{-1} \int_{A_k} P_k (\underline{U}_k - \underline{W}_k) \cdot \underline{n}_k dA$$
$$- v^{-1} \int_{A_k} \rho_k h_k (\underline{U}_k - \underline{W}_k) \cdot \underline{n}_k dA. \qquad (4.6.5)$$

The time-averaged results of some individual terms in Eq. (4.6.5) were given previously; the second, third, and fourth terms in the RHS of Eq. (4.6.5) were given by Eqs. (5.7.4c), (5.9.3e), and (5.7.6d), respectively. The fifth term consists of four items that were given by Eqs. (5.9.5), (5.7.7b), (5.7.7c), and (5.7.7e), respectively. The following describes the time averaging of remaining terms of Eq. (4.6.5):

The first term on the LHS of Eq. (4.6.5) is

$$\gamma_v \frac{\partial}{\partial t} {}^{t}\langle \alpha_k {}^{3i}\langle \rho_k h_k \rangle\rangle \tag{5.11.1}$$

$$\begin{aligned} &{}^{t}\langle \alpha_k {}^{3i}\langle \rho_k h_k \rangle\rangle \\ &= {}^{t}\langle \alpha_k {}^{3i}\langle ({}^{3i}\langle \rho_k \rangle_{LF} + \tilde{\rho}_{kLF} + \rho_k')({}^{3i}\langle h_k \rangle_{LF} + \tilde{h}_{kLF} + h_k')\rangle\rangle \\ &= \alpha_k {}^{3i}\langle \rho_k \rangle_{LF} {}^{3i}\langle h_k \rangle_{LF} + \phi_{hk}^{3i}, \end{aligned} \tag{5.11.1a}$$

in which ϕ_{hk}^{3i} is a scalar enthalpy function defined by

$$\phi_{hk}^{3i} = \alpha_k {}^{3i}\langle \tilde{\rho}_{kLF} \tilde{h}_{kLF} \rangle + \alpha_k {}^{t3i}\langle \rho_k' h_k' \rangle. \tag{5.11.1b}$$

The first term on the LHS of Eq. (4.6.5) then can be written as

$$\begin{aligned} &\gamma_v \frac{\partial}{\partial t} {}^{t}\langle \alpha_k {}^{3i}\langle \rho_k h_k \rangle\rangle \\ &\quad = \gamma_v \frac{\partial}{\partial t} (\alpha_k {}^{3i}\langle \rho_k \rangle_{LF} {}^{3i}\langle h_k \rangle_{LF} + \phi_{hk}^{3i}). \end{aligned} \tag{5.11.1c}$$

The second term on the LHS of Eq. (4.6.5) is

$$\gamma_A \nabla \cdot {}^{t}\langle \alpha_k {}^{2i}\langle \rho_k \underline{U}_k h_k \rangle\rangle \tag{5.11.2}$$

$$\begin{aligned} &{}^{t}\langle \alpha_k {}^{2i}\langle \rho_k \underline{U}_k h_k \rangle\rangle \\ &= {}^{t}\langle \alpha_k {}^{2i}\langle ({}^{2i}\langle \rho_k \rangle_{LF} + \tilde{\rho}_{kLF} + \rho_k')({}^{2i}\langle \underline{U}_k \rangle_{LF} \\ &\quad + \underline{\tilde{U}}_{kLF} + \underline{U}_k')({}^{2i}\langle h_k \rangle_{LF} + \tilde{h}_{kLF} + h_k')\rangle\rangle \\ &= \alpha_k {}^{2i}\langle \rho_k \rangle_{LF} {}^{2i}\langle \underline{U}_k \rangle_{LF} {}^{2i}\langle h_k \rangle_{LF} + {}^{2i}\langle \underline{U}_k \rangle_{LF} \phi_{hk}^{2i} \\ &\quad + {}^{2i}\langle h_k \rangle_{LF} \underline{\psi}_{mk}^{2i} \\ &\quad + \alpha_k ({}^{2i}\langle \underline{h}_k^T \rangle + {}^{2i}\langle \tilde{h}_k \rangle + {}^{2i}\langle \underline{\tilde{h}}_k^T \rangle), \end{aligned} \tag{5.11.2a}$$

where $\phi_{hk}^{2i} = \alpha_k {}^{2i}\langle \tilde{\rho}_{kLF} \tilde{h}_{kLF} \rangle + \alpha_k {}^{t2i}\langle \rho_k' h_k' \rangle$. (5.11.2b)

$\underline{\psi}_{mk}^{2i}$ is defined in Eq. (5.3.2h).

(a) The volume-averaged turbulent enthalpy flux ${}^{2i}\langle \underline{h}_k^T \rangle$ is defined by

$$ {}^{2i}\langle \rho_k \rangle_{LF} {}^{t2i}\langle \underline{U}_k' h_k' \rangle + {}^{t2i}\langle \tilde{\rho}_{kLF} \underline{U}_k' h_k' \rangle. \quad (5.11.2c) $$

(b) The volume-averaged dispersive enthalpy flux ${}^{2i}\langle \underline{\tilde{h}}_k \rangle$ is defined by

$$ {}^{2i}\langle \rho_k \rangle_{LF} {}^{2i}\langle \underline{\tilde{U}}_{kLF} \tilde{h}_{kLF} \rangle + {}^{2i}\langle \tilde{\rho}_{kLF} \underline{\tilde{U}}_{kLF} \tilde{h}_{kLF} \rangle. \quad (5.11.2d) $$

(c) The volume-averaged turbulent, dispersive enthalpy flux ${}^{2i}\langle \underline{\tilde{h}}_k^T \rangle$ is defined by

$$ {}^{t2i}\langle \underline{\tilde{U}}_{kLF} \rho_k' h_k' \rangle + {}^{t2i}\langle \underline{U}_k' \rho_k' \tilde{h}_k \rangle. \quad (5.11.2e) $$

(d) $\rho_k' \underline{U}_k' h_k'$ is a triple time correlation that is assumed to be small and can be neglected. (5.11.2f)

The second term on the LHS of Eq. (4.6.5) can be written as follows.

$$ \begin{aligned} &\gamma_A \nabla \cdot {}^t\langle \alpha_k {}^{2i}\langle \rho_k \underline{U}_k h_k \rangle \rangle \\ &= \gamma_A \nabla \cdot \left[\alpha_k {}^{2i}\langle \rho_k \rangle_{LF} {}^{2i}\langle \underline{U}_k \rangle_{LF} {}^{2i}\langle h_k \rangle_{LF} + {}^{2i}\langle \underline{U}_k \rangle_{LF} \phi_{hk}^{2i} \right. \\ &\quad \left. + {}^{2i}\langle h_k \rangle_{LF} \underline{\psi}_{mk}^{2i} + \alpha_k \left({}^{2i}\langle \underline{h}_k^T \rangle + {}^{2i}\langle \underline{\tilde{h}}_k \rangle + {}^{2i}\langle \underline{\tilde{h}}_k^T \rangle \right) \right]. \end{aligned} \quad (5.11.2g) $$

The first term on the RHS of Eq. (4.6.5) is

$$ \gamma_v \frac{\partial}{\partial t} {}^t\langle \alpha_k {}^{3i}\langle P_k \rangle \rangle \quad (5.11.3) $$

$$ \begin{aligned} {}^t\langle \alpha_k {}^{3i}\langle P_k \rangle \rangle &= {}^t\langle \alpha_k {}^{3i}\langle {}^{3i}\langle P_k \rangle_{LF} + \tilde{P}_{kLF} + P_k' \rangle \rangle \\ &= {}^t\langle \alpha_k ({}^{3i}\langle P_k \rangle_{LF} + P') \rangle \\ &= \alpha_k {}^{3i}\langle P_k \rangle_{LF}. \end{aligned} \quad (5.11.3a) $$

The first term on the RHS of Eq. (4.6.5) can be written as

$$\gamma_v \frac{\partial}{\partial t} \alpha_k {}^{3i}\langle P_k \rangle_{LF}. \tag{5.11.3b}$$

The sixth term on the RHS of Eq. (4.6.5) is

$$\begin{aligned} &{}^t\left\langle v^{-1} \int_{A_k} P_k \left(\underline{U}_k - \underline{W}_k\right) \cdot \underline{n}_k dA \right\rangle \\ &\quad = v^{-1} \int_{A_k} {}^t\langle P_k \left(\underline{U}_k - \underline{W}_k\right) \cdot \underline{n}_k dA \rangle \end{aligned} \tag{5.11.8}$$

$$\begin{aligned} &v^{-1} \int_{A_k} {}^t\langle P_k(\underline{U}_k - \underline{W}_k) \cdot \underline{n}_k dA \rangle \\ &\quad = v^{-1} \int_{A_k} {}^t\langle ({}^{3i}\langle P_k \rangle_{LF} + \tilde{P}_{kLF} + P'_k)[({}^{3i}\langle \underline{U}_k \rangle_{LF} \\ &\qquad + \underline{\tilde{U}}_{kLF} + \underline{U}'_k) - (\underline{W}_{kLF} + \underline{W}'_k)] \cdot \underline{n}_k dA \rangle. \end{aligned} \tag{5.11.8a}$$

Because low- and high-frequency parts of Eq. (5.11.6a) can be separated, after time averaging the equation becomes

$$\begin{aligned} &v^{-1} \int_{A_k} {}^t\langle P_k \left(\underline{U}_k - \underline{W}_k\right) \cdot \underline{n}_k dA \rangle \\ &= -\gamma_v {}^{3i}\langle P_k \rangle_{LF} \left(\frac{\partial \alpha_k}{\partial t} + {}^{3i}\langle \underline{U}_k \rangle_{LF} \cdot \nabla \alpha_k\right) - (PWI)_k^{(u)} + (PWI)_k^{(h)} \\ &= -\gamma_v {}^{3i}\langle P_k \rangle_{LF} \frac{d\alpha_k}{dt_k} - (PWI)_k^{(u)} + (PWI)_k^{(h)}, \end{aligned} \tag{5.11.8b}$$

$$\text{where } (PWI)_k^{(u)} = -v^{-1}\, {}^{3i}\langle P_k \rangle_{LF} \int_{A_k} \underline{\tilde{U}}_{kLF} \cdot \underline{n}_k dA \tag{5.9.3d}$$

$$\begin{aligned} (PWI)_k^{(h)} &= v^{-1} \int_{A_k} \tilde{P}_{kLF}({}^{3i}\langle \underline{U}_k \rangle_{LF} + \underline{\tilde{U}}_{kLF} - \underline{W}_{kLF}) \\ &\quad \cdot \underline{n}_k dA + v^{-1} \int_{A_k} {}^t\langle P'_k \left(\underline{U}'_k - \underline{W}'_k\right)\rangle \cdot \underline{n}_k dA. \end{aligned} \tag{5.11.8c}$$

The sixth term on the RHS of Eq. (4.6.5) thus can be written as

$$-\gamma_v {}^{3i}\langle P_k\rangle_{LF}\frac{d\alpha_k}{dt_k}-(PWI)_k^{(u)}+(PWI)_k^{(h)}. \quad (5.11.8d)$$

The seventh term on the RHS of Eq. (4.6.5) is

$$^{t}\left\langle -v^{-1}\int_{A_k}\rho_k h_k\left(\underline{U}_k-\underline{W}_k\right)\cdot\underline{n}_k dA\right\rangle = -v^{-1}\int_{A_k} {}^{t}\langle\rho_k h_k\left(\underline{U}_k-\underline{W}_k\right)\cdot\underline{n}_k dA\rangle \quad (5.11.9)$$

$$-v^{-1}\int_{A_k} {}^{t}\langle\rho_k h_k\left(\underline{U}_k-\underline{W}_k\right)\cdot\underline{n}_k dA\rangle$$
$$= -v^{-1}\int_{A_k} {}^{t}\langle({}^{3i}\langle\rho_k\rangle_{LF}+\tilde{\rho}_{kLF}+\rho_k')({}^{3i}\langle h_k\rangle_{LF}+\tilde{h}_{kLF}$$
$$+h_k')\rangle[({}^{3i}\langle\underline{U}_k\rangle_{LF}+\underline{\tilde{U}}_{kLF}+\underline{U}_k')-(\underline{W}_{kLF}+\underline{W}_k')]\cdot\underline{n}_k dA. \quad (5.11.9a)$$

Because the low- and high-frequency parts of Eq. (5.11.9a) can be separated after time averaging, the equation becomes

$$-v^{-1}\int_{A_k} {}^{t}\langle\rho_k h_k(\underline{U}_k-\underline{W}_k)\rangle\cdot\underline{n}_k dA$$
$$=\gamma_v {}^{3i}\langle\rho_k\rangle_{LF} {}^{3i}\langle h_k\rangle_{LF}\left(\frac{\partial\alpha_k}{\partial t}+{}^{3i}\langle\underline{U}_k\rangle_{LF}\cdot\nabla\alpha_k\right)$$
$$-v^{-1}\,{}^{3i}\langle\rho_k\rangle_{LF} {}^{3i}\langle h_k\rangle_{LF}\int_{A_k}\underline{\tilde{U}}_{kLF}\cdot\underline{n}_k dA$$
$$-v^{-1}\,{}^{3i}\langle h_k\rangle_{LF}\int_{A_k}\tilde{\rho}_{kLF}({}^{3i}\langle\underline{U}_k\rangle_{FL}+\underline{\tilde{U}}_{kLF}-\underline{W}_{kLF})\cdot\underline{n}_k dA$$
$$-v^{-1}\int_{A_k}({}^{3i}\langle\rho_k\rangle_{LF}+\tilde{\rho}_{kLF})\tilde{h}_{kLF}({}^{3i}\langle\underline{U}_k\rangle_{LF}$$
$$+\underline{\tilde{U}}_{kLF}-\underline{W}_{kLF})\cdot\underline{n}_k dA$$

$$- \upsilon^{-1} \int_{A_k} ({}^{3i}\langle \rho_k \rangle_{LF} + \tilde{\rho}_{kLF}) {}^t\langle h'_k (\underline{U}'_k - \underline{W}'_k) \rangle \cdot \underline{n}_k dA$$

$$- \upsilon^{-1} \int_{A_k} ({}^{3i}\langle h_k \rangle_{LF} + \tilde{h}_{kLF}) {}^t\langle \rho'_k (\underline{U}'_k - \underline{W}'_k) \rangle \cdot \underline{n}_k dA$$

$$- \upsilon^{-1} {}^t\left\langle \int_{A_k} ({}^{3i}\langle \underline{U}_k \rangle_{LF} + \underline{\tilde{U}}_{kLF} - \underline{W}_{kLF}) {}^t\langle \rho'_k h'_k \rangle \cdot \underline{n}_k dA \right\rangle$$

$$- \upsilon^{-1} \int_{A_k} {}^t\langle \rho'_k h'_k (\underline{U}'_k - \underline{W}'_k) \rangle \cdot \underline{n}_k dA. \qquad (5.11.9b)$$

The seventh term on the RHS of Eq. (4.6.5) thus can be written by multiplying Eq. (5.3.3c) by ${}^{3i}\langle h_k \rangle_{LF}$. Introducing the results into Eq. (5.11.9b) leads to

$$-\upsilon^{-1} \int_{A_k} {}^t\langle \rho_k h_k (\underline{U}_k - \underline{W}_k) \rangle \cdot \underline{n}_k dA$$
$$= \gamma_\upsilon \alpha_k {}^t\langle \Gamma_k \rangle {}^{3i}\langle h_k \rangle_{LF} + (EPYTI)_k. \qquad (5.11.9c)$$

The interfacial enthalpy transfer integral is

$$(EPYTI)_k$$
$$= -\upsilon^{-1} \int_{A_k} \tilde{h}_{kLF} ({}^{3i}\langle \rho_k \rangle_{LF} + \tilde{\rho}_{kLF})$$
$$\times ({}^{3i}\langle \underline{U}_k \rangle_{LF} + \underline{\tilde{U}}_{kLF} - \underline{W}_{kLF}) \cdot \underline{n}_k dA$$
$$- \upsilon^{-1} \int_{A_k} ({}^{3i}\langle \rho_k \rangle_{LF} + \tilde{\rho}_{kLF}) {}^t\langle h'_k (\underline{U}'_k - \underline{W}'_k) \rangle \cdot \underline{n}_k dA$$
$$- \upsilon^{-1} \int_{A_k} \tilde{h}_{kLF} {}^t\langle \rho'_k (\underline{U}'_k - \underline{W}'_k) \rangle \cdot \underline{n}_k dA$$
$$- \upsilon^{-1} \int_{A_k} ({}^{3i}\langle \underline{U}_k \rangle_{LF} + \underline{\tilde{U}}_{kLF} - \underline{W}_{kLF}) {}^t\langle \rho'_k h'_k \rangle \cdot \underline{n}_k dA$$
$$- \upsilon^{-1} \int_{A_k} {}^t\langle \rho'_k h'_k (\underline{U}'_k - \underline{W}'_k) \rangle \cdot \underline{n}_k dA. \qquad (5.11.9d)$$

The time-averaging interfacial enthalpy source of phase k per unit volume in υ is very similar to the corresponding equation of internal energy

$$\gamma_\upsilon \alpha_k {}^t\langle \Pi_k \rangle = \gamma_\upsilon \alpha_k ({}^{t3i}\langle \dot{Q}_{kf}\rangle + {}^{t3i}\langle \dot{Q}_{wk}\rangle) - \gamma_\upsilon {}^{3i}\langle P_k\rangle_{LF} \frac{d\alpha_k}{dt_k} + (PWI)_k^{(h)} + (VDI)_k + \gamma_\upsilon \alpha_k {}^t\langle \Gamma_k\rangle {}^{3i}\langle h_k\rangle_{LF} + (EPYTI)_k, \quad (5.11.9e)$$

in which the substantive time derivative $\frac{d\alpha_k}{dt_k}$ is defined in Eq. (5.3.3e). We note that the term $-(PWI)_k^{(u)}$ arising from the time averaging of $\upsilon^{-1}\int_{A_k} {}^t\langle P_k(\underline{U}_k - \underline{W}_k)\rangle \cdot \underline{n}_k dA$ in Eq. (5.11.8b) canceled with $+(PWI)_k^{(u)}$ arising from the time average of $-\gamma_\upsilon \alpha_k {}^{3i}\langle P_k \nabla \cdot \underline{U}_k\rangle$, as shown in Eq. (5.9.3e). The term $(VDI)_k$ arises from the time average of $\alpha_k {}^t\langle \Phi_k\rangle$, as shown in Eq. (5.9.5).

By using results of Eqs. (5.11.1c), (5.11.2g), (5.11.3b), (5.7.4c), (5.9.3e), (5.7.6d), (5.9.5), (5.7.7b), (5.7.7c), (5.7.7e), (5.11.8b), and (5.11.9c), we obtain the time volume-averaged enthalpy equation as follows:

$$\begin{aligned}
&\gamma_\upsilon \frac{\partial}{\partial t}(\alpha_k {}^{3i}\langle \rho_k\rangle_{LF} {}^{3i}\langle h_k\rangle_{LF} + \phi_{hk}^{3i}) \\
&\quad + \gamma_A \nabla \cdot [\alpha_k {}^{2i}\langle \rho_k\rangle_{LF} {}^{2i}\langle \underline{U}_k\rangle_{LF} {}^{2i}\langle h_k\rangle_{LF} + {}^{2i}\langle \underline{U}_k\rangle_{LF} \phi_{hk}^{2i} \\
&\quad + {}^{2i}\langle h_k\rangle_{LF} \underline{\psi}_{mk}^{2i} + \alpha_k({}^{2i}\langle \underline{h}_k^T\rangle + {}^{2i}\langle \underline{\tilde{h}}_k\rangle + {}^{2i}\langle \underline{\tilde{h}}_k^T\rangle)] \\
&= \gamma_\upsilon \frac{\partial}{\partial t}\alpha_k {}^{3i}\langle P_k\rangle_{LF} + \gamma_A \nabla \cdot (\alpha_k {}^{2i}\langle P_k\rangle_{LF} {}^{2i}\langle \underline{U}_k\rangle_{LF} + \underline{\Psi}_{Pk}^{2i}) \\
&\quad - \gamma_\upsilon(\alpha_k {}^{3i}\langle P_k\rangle_{LF} \nabla \cdot {}^{3i}\langle \underline{U}_k\rangle_{LF} + \phi_{Pk}^{3i}) \\
&\quad - \gamma_A \nabla \cdot \alpha_k {}^{2i}\langle \underline{J}_{qk}\rangle_{LF} \\
&\quad + \gamma_\upsilon(\alpha_k {}^{3i}\langle \underline{\underline{\tau}}_k\rangle_{LF} : \nabla, {}^{3i}\langle \underline{U}_k\rangle_{LF} + \phi_{\tau k}^{3i}) + (VDI)_k
\end{aligned}$$

$$+\gamma_v \alpha_k \left({}^{3i}\langle J_{Ek}\rangle_{LF} + {}^{t3i}\langle \dot{Q}_{kf}\rangle + {}^{t3i}\langle \dot{Q}_{wf}\rangle \right)$$
$$-\gamma_v {}^{3i}\langle P_k\rangle_{LF} \frac{d\alpha_k}{dt_k} + (PWI)_k^{(h)}$$
$$+\gamma_v {}^t\alpha_k \langle \Gamma_k\rangle {}^{3i}\langle h_k\rangle_{LF} + (EPYTI)_k . \qquad (5.11.10)$$

When ρ_k = constant, ${}^{3i}\langle \rho_k\rangle_{LF} = \rho_k$, $\tilde{\rho}_{kLF} = \rho'_k = 0$. Therefore,

$$\phi_{hk}^{3i} = 0,\ \phi_{hk}^{2i} = 0,\ \underline{\psi}_{mk}^{3i} = 0 \text{ and } \underline{\psi}_{mk}^{2i} = 0 \qquad (5.11.11)$$

$$ {}^{o2i}\langle \underline{h}_k^T\rangle = \rho_k {}^{t2i}\langle \underline{U}'_k h'_k\rangle \qquad (5.11.11a)$$

$$ {}^{o2i}\langle \underline{\tilde{h}}_k\rangle = \rho_k {}^{2i}\langle \underline{\tilde{U}}_{kLF} \tilde{h}_{kLF}\rangle \qquad (5.11.11b)$$

$$ {}^{o2i}\langle \underline{\tilde{h}}_k^T\rangle = 0 \qquad (5.11.11c)$$

$$\gamma_v \alpha_k {}^{ot}\langle \Gamma_k\rangle = \gamma_v \rho_k \left(\frac{\partial \alpha_k}{\partial t} + {}^{3i}\langle \underline{U}_k\rangle_{LF} \cdot \nabla \alpha_k \right)$$
$$+ {}^o\langle MTI\rangle_k \qquad (5.11.11d)$$

$$ {}^o\langle MTI\rangle_k = -v^{-1}\rho_k \int_{A_k} \underline{\tilde{U}}_{kLF} \cdot \underline{n}_k dA \qquad (5.11.11e)$$

$$ {}^o(EPYTI)_k$$
$$= -v^{-1}\rho_k \int_{A_k} \tilde{h}_{kLF} ({}^{3i}\langle \underline{U}_k\rangle_{LF} + \underline{\tilde{U}}_{kLF} - \underline{W}_{kLF}) \cdot \underline{n}_k dA$$
$$-v^{-1}\rho_k \int_{A_k} h'_k (\underline{U}'_k - \underline{W}'_k) \cdot \underline{n}_k dA. \qquad (5.11.11f)$$

The time volume-averaged enthalpy equation for ρ_k = constant can be written as follows:

$$\gamma_v \rho_k \frac{\partial}{\partial t} (\alpha_k {}^{3i}\langle h_k\rangle_{LF}) + \gamma_A \nabla \cdot \left[\alpha_k \rho_k {}^{2i}\langle \underline{U}_k\rangle_{LF} {}^{2i}\langle h_k\rangle_{LF} \right.$$
$$\left. + \alpha_k \left({}^{o2i}\langle \underline{h}_k^T\rangle + {}^{o2i}\langle \underline{\tilde{h}}_k\rangle \right) \right]$$

$$
\begin{aligned}
&= \gamma_v \frac{\partial}{\partial t} \alpha_k{}^{3i}\langle P_k\rangle_{LF} + \gamma_A \nabla \cdot (\alpha_k{}^{2i}\langle \underline{U}_k\rangle_{LF}{}^{2i}\langle P_k\rangle_{LF} + \underline{\Psi}^{2i}_{Pk}) \\
&- \gamma_v (\alpha_k{}^{3i}\langle P_k\rangle_{LF} \nabla \cdot {}^{3i}\langle \underline{U}_k\rangle_{LF} + \phi^{3i}_{Pk}) \\
&- \gamma_A \nabla \cdot \alpha_k{}^{2i}\langle \underline{J}_{qk}\rangle_{LF} \\
&+ \gamma_v (\alpha_k{}^{3i}\langle \underline{\underline{\tau}}_k\rangle_{LF} : \nabla, {}^{3i}\langle \underline{U}_k\rangle_{LF} + \phi^{3i}_{\tau k}) + (VDI)_k \\
&+ \gamma_v \alpha_k ({}^{3i}\langle J_{Ek}\rangle_{LF} + {}^{t3i}\langle \dot{Q}_{kf}\rangle + {}^{t3i}\langle \dot{Q}_{wk}\rangle) \\
&- \gamma_v {}^{3i}\langle P_k\rangle_{LF} \frac{d\alpha_k}{dt_k} + (PWI)^{(h)}_k \\
&+ \gamma_v \alpha_k{}^{ot}\langle \Gamma_k\rangle^{3i}\langle h_k\rangle_{LF} + {}^{o}(EPYTI)_k .
\end{aligned}
\tag{5.11.12}
$$

5.12 Time-volume-averaged interfacial enthalpy balance equation (capillary energy ignored)

The local volume-averaged enthalpy balance equation is given by Eq. (4.6.5), the time volume-averaged interfacial enthalpy balance equation can readily be written by using Eq. (5.11.9e):

$$
\begin{aligned}
&\gamma_v \alpha_k ({}^{t3i}\langle \dot{Q}_{kf}\rangle + {}^{t3i}\langle \dot{Q}_{wk}\rangle) - \gamma_v {}^{3i}\langle P_k\rangle_{LF} \frac{d\alpha_k}{dt_k} + (PWI)^{(h)}_k \\
&+ (VDI)_k + \gamma_v \alpha_k{}^{t}\langle \Gamma_k\rangle^{3i}\langle h_k\rangle_{LF} + (EPYTI)_k \\
&+ \gamma_v \alpha_f ({}^{t3i}\langle \dot{Q}_{fk}\rangle + {}^{t3i}\langle \dot{Q}_{wf}\rangle) - \gamma_v {}^{3i}\langle P_f\rangle_{LF} \frac{d\alpha_f}{dt_f} \\
&+ (PWI)^{(h)}_f + (VDI)_f + \gamma_v \alpha_f{}^{t}\langle \Gamma_f\rangle^{3i}\langle h_f\rangle_{LF} \\
&+ (EPYTI)_f = 0.
\end{aligned}
\tag{5.12.1}
$$

Here, $(PWI)^{(h)}_k$ and $(EPYTI)_k$ are defined in Eqs. (5.11.8c) and (5.11.9d). Similarly for phase f, we can write

$$
\begin{aligned}
(PWI)^{(h)}_f &= \upsilon^{-1} \int_{A_f} \tilde{P}_{fLF} ({}^{3i}\langle \underline{U}_f\rangle_{LF} + \underline{\tilde{U}}_{fLF} - \underline{W}_{fLF}) \cdot \underline{n}_f dA \\
&+ \upsilon^{-1} \int_{A_f} {}^{t}\langle P'_f (\underline{U}'_f - \underline{W}'_f)\rangle \cdot \underline{n}_f dA
\end{aligned}
\tag{5.12.2}
$$

$$
\begin{aligned}
&(EPYTI)_f \\
&= -\upsilon^{-1} \int_{A_f} \tilde{h}_{fLF} ({}^{3i}\langle \rho_f \rangle_{LF} + \tilde{\rho}_{fLF}) \\
&\quad \times ({}^{3i}\langle \underline{U}_f \rangle_{LF} + \underline{\tilde{U}}_{fLF} - \underline{W}_{fLF}) \cdot \underline{n}_f dA \\
&\quad - \upsilon^{-1} \int_{A_f} ({}^{3i}\langle \rho_f \rangle_{LF} + \tilde{\rho}_{fLF})^t \langle h'_f (\underline{U}'_f - \underline{W}'_f) \rangle \cdot \underline{n}_f dA \\
&\quad - \upsilon^{-1} \int_{A_f} \tilde{h}_{fLF}{}^t \langle \rho'_f (\underline{U}'_f - \underline{W}'_f) \rangle \cdot \underline{n}_f dA \\
&\quad - \upsilon^{-1} \int_{A_f} ({}^{3i}\langle \underline{U}_f \rangle_{LF} + \underline{\tilde{U}}_{fLF} - \underline{W}_{fLF})^t \langle \rho'_f h'_f \rangle \cdot \underline{n}_f dA \\
&\quad - \upsilon^{-1} \int_{A_f} {}^t \langle \rho'_f h'_f (\underline{U}'_f - \underline{W}'_f) \rangle \cdot \underline{n}_f dA. \qquad (5.12.3)
\end{aligned}
$$

It is to be noted that both enthalpy dissipation and enthalpy production due to interfacial pressure work integral are shown in interfacial enthalpy balance equation.

5.13 Summary of time-volume-averaged conservation equations

The time-volume-averaged conservation of mass, momentum, and energy (total energy, internal energy, and enthalpy) equations are listed this section.

5.13.1 Time-volume-averaged conservation of mass equation

$$
\begin{aligned}
\gamma_\upsilon \frac{\partial}{\partial t} \alpha_k {}^{3i}\langle \rho_k \rangle_{LF} &+ \gamma_A \nabla \cdot \alpha_k {}^{2i}\langle \rho_k \rangle_{LF} {}^{2i}\langle \underline{U}_k \rangle_{LF} \\
&+ \gamma_A \nabla \cdot \underline{\psi}^{2i}_{mk} = \gamma_\upsilon \alpha_k {}^t\langle \Gamma_k \rangle \qquad (5.3.4)
\end{aligned}
$$

$$\gamma_v \alpha_k {}^t\langle \Gamma_k \rangle = -\upsilon^{-1} \int_{A_k} {}^t\langle \rho_k (\underline{U}_k - \underline{W}_k)\rangle \cdot \underline{n}_k dA$$
$$= \gamma_v {}^{3i}\langle \rho_k \rangle_{LF} \left(\frac{\partial \alpha_k}{\partial t} + {}^{3i}\langle \underline{U}_k \rangle_{LF} \cdot \nabla \alpha_k \right)$$
$$+ (MTI)_k \tag{5.3.3c}$$

Here, $(MTI)_k$ is defined in Eq. (5.3.3d).

5.13.2 Time-volume-averaged linear momentum conservation equation

Various forms of the time-volume-averaged linear momentum equations are listed as follows:

$$\gamma_v \frac{\partial}{\partial t} \alpha_k {}^{3i}\langle \rho_k \rangle_{LF} {}^{3i}\langle \underline{U}_k \rangle_{LF} + \gamma_v \frac{\partial}{\partial t} \underline{\psi}^{3i}_{mk}$$
$$+ \gamma_A \nabla \cdot [\alpha_k {}^{2i}\langle \rho_k \rangle_{LF} {}^{2i}\langle \underline{U}_k \rangle_{LF} {}^{2i}\langle \underline{U}_k \rangle_{LF} + 2\, {}^{2i}\langle \underline{U}_k \rangle_{LF} \underline{\psi}^{2i}_{mk}$$
$$- \alpha_k ({}^{2i}\langle \underline{\underline{\tau}}^T_k \rangle + {}^{2i}\langle \underline{\underline{\tilde{\tau}}}_k \rangle + {}^{2i}\langle \underline{\underline{\tilde{\tau}}}^T_k \rangle)] = -\gamma_v \nabla \alpha_k {}^{3i}\langle P_k \rangle_{LF}$$
$$+ \gamma_A \nabla \cdot \alpha_k {}^{2i}\langle \underline{\underline{\tau}}_k \rangle_{LF} + \gamma_v \alpha_k {}^{3i}\langle \rho_k \rangle_{LF} \underline{f}$$
$$+ \gamma_v {}^{3i}\langle P_k \rangle_{LF} \nabla \alpha_k - \gamma_v {}^{3i}\langle \underline{\underline{\tau}}_k \rangle_{LF} \cdot \nabla \alpha_k + (PTI)_k - (VSTI)_k$$
$$+ \gamma_v \alpha_k {}^t\langle \Gamma_k \rangle {}^{3i}\langle \underline{U}_k \rangle_{LF} + (MMTI)_k \tag{5.5.7f}$$

$$\gamma_v \frac{\partial}{\partial t} (\alpha_k {}^{3i}\langle \rho_k \rangle_{LF} {}^{3i}\langle \underline{U}_k \rangle_{LF} + \underline{\psi}^{3i}_{mk})$$
$$+ \gamma_A \nabla \cdot [\alpha_k {}^{2i}\langle \rho_k \rangle_{LF} {}^{2i}\langle \underline{U}_k \rangle_{LF} {}^{2i}\langle \underline{U}_k \rangle_{LF} + 2\, {}^{2i}\langle \underline{U}_k \rangle_{LF} \underline{\psi}^{2i}_{mk}$$
$$- \alpha_k ({}^{2i}\langle \underline{\underline{\tau}}^T_k \rangle + {}^{2i}\langle \underline{\underline{\tilde{\tau}}}_k \rangle + {}^{2i}\langle \underline{\underline{\tilde{\tau}}}^T_k \rangle)] = -\gamma_v \alpha_k \nabla {}^{3i}\langle P_k \rangle_{LF}$$
$$+ \gamma_A \nabla \cdot \alpha_k {}^{2i}\langle \underline{\underline{\tau}}_k \rangle_{LF} + \gamma_v \alpha_k {}^{3i}\langle \rho_k \rangle_{LF} \underline{f}$$
$$- \gamma_v {}^{3i}\langle \underline{\underline{\tau}}_k \rangle_{LF} \cdot \nabla \alpha_k + (PTI)_k - (VSTI)_k$$
$$+ \gamma_v \alpha_k {}^t\langle \Gamma_k \rangle {}^{3i}\langle \underline{U}_k \rangle_{LF} + (MMTI)_k \tag{5.5.7g}$$

$$\gamma_v \frac{\partial}{\partial t}(\alpha_k{}^{3i}\langle\rho_k\rangle_{LF}{}^{3i}\langle\underline{U}_k\rangle_{LF} + \underline{\psi}^{3i}_{mk})$$
$$+\gamma_A \nabla\cdot[\alpha_k{}^{2i}\langle\rho_k\rangle_{LF}{}^{2i}\langle\underline{U}_k\rangle_{LF}{}^{2i}\langle\underline{U}_k\rangle_{LF} + 2\,{}^{2i}\langle\underline{U}_k\rangle_{LF}\underline{\psi}^{2i}_{mk}$$
$$-\alpha_k({}^{2i}\langle\underline{\underline{\tau}}^T_k\rangle + {}^{2i}\langle\underline{\underline{\tilde{\tau}}}_k\rangle + {}^{2i}\langle\underline{\underline{\tilde{\tau}}}^T_k\rangle)]$$
$$= -\gamma_v \nabla\alpha_k{}^{3i}\langle P_k\rangle_{LF} + \gamma_A \nabla\cdot\alpha_k{}^{2i}\langle\underline{\underline{\tau}}_k\rangle_{LF}$$
$$+\gamma_v\alpha_k({}^{3i}\langle\rho_k\rangle_{LF}\underline{f} - {}^{3i}\langle\underline{R}_k\rangle)$$
$$+v^{-1}\int_{A_{kf}} {}^t\langle -P_k\underline{\underline{I}} + \underline{\underline{\tau}}_k\rangle\cdot\underline{n}_k dA$$
$$+\gamma_v\alpha_k{}^t\langle\Gamma_k\rangle{}^{3i}\langle\underline{U}_k\rangle_{LF} + (MMTI)_k. \tag{5.5.7j}$$

In the preceding three equations, $(PTI)_k$, $(VSTI)_k$, and $(MMTI)_k$ are given in Eqs. (5.5.6c), (5.5.6d), and (5.5.7d), respectively.

5.13.3 Time-volume-averaged total energy conservation equation

$$\gamma_v \frac{\partial}{\partial t}(\alpha_k{}^{3i}\langle\rho_k\rangle_{LF}{}^{3i}\langle E_k\rangle_{LF} + \phi^{3i}_{Ek})$$
$$+\gamma_A \nabla\cdot[\alpha_k{}^{2i}\langle\rho_k\rangle_{LF}{}^{2i}\langle\underline{U}_k\rangle_{LF}{}^{2i}\langle E_k\rangle_{LF} + \phi^{2i}_{Ek}{}^{2i}\langle\underline{U}_k\rangle_{LF}$$
$$+\underline{\psi}^{2i}_{mk}{}^{2i}\langle E_k\rangle_{LF} + \alpha_k({}^{2i}\langle\underline{E}^T_k\rangle + {}^{2i}\langle\underline{\tilde{E}}_k\rangle + {}^{2i}\langle\underline{\tilde{E}}^T_k\rangle)]$$
$$= -\gamma_A \nabla\cdot(\alpha_k{}^{2i}\langle\underline{U}_k\rangle_{LF}{}^{2i}\langle P_k\rangle_{LF} + \underline{\Psi}^{2i}_{Pk})$$
$$+\gamma_A \nabla\cdot(\alpha_k{}^{2i}\langle\underline{U}_k\rangle_{LF}\cdot{}^{2i}\langle\underline{\underline{\tau}}_k\rangle_{LF} + \underline{\Psi}^{2i}_{\tau k})$$
$$-\gamma_A \nabla\cdot\alpha_k{}^{2i}\langle\underline{J}_{qk}\rangle_{LF} + \gamma_v[(\alpha_k{}^{3i}\langle\rho_k\rangle_{LF}{}^{3i}\langle\underline{U}_k\rangle_{LF}\cdot\underline{f}$$
$$+\underline{\psi}^{3i}_{mk}\cdot\underline{f}) + \alpha_k{}^{3i}\langle J_{Ek}\rangle_{LF} + \alpha_k{}^{t3i}\langle\dot{Q}_{kf}\rangle$$
$$+\alpha_k{}^{t3i}\langle\dot{Q}_{wk}\rangle] + \gamma_v{}^{3i}\langle\underline{U}_k\rangle_{LF}{}^{3i}\langle P_k\rangle_{LF}\cdot\nabla\alpha_k$$
$$-\gamma_v{}^{3i}(\langle\underline{U}_k\rangle_{LF}\cdot{}^{3i}\langle\underline{\underline{\tau}}_k\rangle_{LF})\cdot\nabla\alpha_k + (PWI)_k$$
$$-(VWI)_k + \gamma_v\alpha_k{}^t\langle\Gamma_k\rangle{}^{3i}\langle E_k\rangle_{LF} + (TETI)_k \tag{5.7.10}$$

Here, the definitions of $(PWI)_k$, $(VWI)_k$, and $(TETI)_k$ are defined in Eqs. (5.7.8b), (5.7.8c), and (5.7.9d).

5.13.4 Time-volume-averaged internal energy conservation equation

$$
\begin{aligned}
&\gamma_v \frac{\partial}{\partial t}\left(\alpha_k\, {}^{3i}\langle\rho_k\rangle_{LF}\, {}^{3i}\langle u_k\rangle_{LF} + \phi^{3i}_{uk}\right) \\
&\quad + \gamma_A \nabla\cdot\left[\alpha_k\, {}^{2i}\langle\rho_k\rangle_{LF}\, {}^{2i}\langle\underline{U}_k\rangle_{LF}\, {}^{2i}\langle u_k\rangle_{LF} + \phi^{2i}_{uk}\, {}^{2i}\langle\underline{U}_k\rangle_{LF}\right. \\
&\quad \left. + \underline{\psi}^{2i}_{mk}\, {}^{2i}\langle u_k\rangle_{LF} + \alpha_k\left({}^{2i}\langle\underline{u}^T_k\rangle + {}^{2i}\langle\underline{\tilde{u}}_k\rangle + {}^{2i}\langle\underline{\tilde{u}}^T_k\rangle\right)\right] \\
&= -\gamma_v\left(\alpha_k\, {}^{3i}\langle P_k\rangle_{LF} \nabla\cdot {}^{3i}\langle\underline{U}_k\rangle_{LF} + \phi^{3i}_{Pk}\right) \\
&\quad + (PWI)^{(u)}_k - \gamma_A \nabla\cdot\alpha_k\, {}^{2i}\langle\underline{J}_{qk}\rangle_{LF} \\
&\quad + \gamma_v\left(\alpha_k\, {}^{3i}\langle\underline{\underline{\tau}}_k\rangle_{LF} : \nabla,\, {}^{3i}\langle\underline{U}_k\rangle_{LF} + \phi^{3i}_{\tau k}\right) + (VDI)_k \\
&\quad + \gamma_v\alpha_k\left({}^{3i}\langle J_{Ek}\rangle_{LF} + {}^{t3i}\langle\dot{Q}_{kf}\rangle + {}^{t3i}\langle\dot{Q}_{wk}\rangle\right) \\
&\quad + \gamma_v\alpha_k\, {}^{t}\langle\Gamma_k\rangle^{3i}\langle u_k\rangle_{LF} + (IETI)_k. \qquad (5.9.7)
\end{aligned}
$$

The definitions of $(PWI)^{(u)}_k$, $(VDI)_k$, and $(IETI)_k$ are shown in Eqs. (5.9.3d), (5.9.5b), and (5.9.6d).

5.13.5 Time-volume-averaged enthalpy conservation equation

$$
\begin{aligned}
&\gamma_v \frac{\partial}{\partial t}\left(\alpha_k\, {}^{3i}\langle\rho_k\rangle_{LF}\, {}^{3i}\langle h_k\rangle_{LF} + \phi^{3i}_{hk}\right) \\
&\quad + \gamma_A \nabla\cdot\left[\alpha_k\, {}^{2i}\langle\rho_k\rangle_{LF}\, {}^{2i}\langle\underline{U}_k\rangle_{LF}\, {}^{2i}\langle h_k\rangle_{LF} + {}^{2i}\langle\underline{U}_k\rangle_{LF}\phi^{2i}_{hk}\right. \\
&\quad \left. + {}^{2i}\langle h_k\rangle_{LF}\underline{\psi}^{2i}_{mk} + \alpha_k\left({}^{2i}\langle\underline{h}^T_k\rangle + {}^{2i}\langle\underline{\tilde{h}}_k\rangle + {}^{2i}\langle\underline{\tilde{h}}^T_k\rangle\right)\right]
\end{aligned}
$$

$$= \gamma_v \frac{\partial}{\partial t}\alpha_k {}^{3i}\langle P_k\rangle_{LF} + \gamma_A \nabla \cdot (\alpha_k {}^{2i}\langle P_k\rangle_{LF} {}^{2i}\langle \underline{U}_k\rangle_{LF} + \underline{\Psi}^{2i}_{Pk})$$

$$- \gamma_v (\alpha_k {}^{3i}\langle P_k\rangle_{LF} \nabla \cdot {}^{3i}\langle \underline{U}_k\rangle_{LF} + \phi^{3i}_{Pk}) - \gamma_A \nabla \cdot \alpha_k {}^{2i}\langle \underline{J}_{qk}\rangle_{LF}$$

$$+ \gamma_v (\alpha_k {}^{3i}\langle \underline{\underline{\tau}}_k\rangle_{LF} : \nabla, {}^{3i}\langle \underline{U}_k\rangle_{LF} + \phi^{3i}_{\tau k}) + (VDI)_k$$

$$+ \gamma_v \alpha_k \left({}^{3i}\langle J_{Ek}\rangle_{LF} + {}^{t3i}\langle \dot{Q}_{kf}\rangle + {}^{t3i}\langle \dot{Q}_{wf}\rangle \right)$$

$$- \gamma_v {}^{3i}\langle P_k\rangle_{LF} \frac{d\alpha_k}{dt_k} + (PWI)^{(h)}_k$$

$$+ \gamma_v \alpha_k {}^{t}\langle \Gamma_k\rangle^{3i}\langle h_k\rangle_{LF} + (EPYTI)_k \qquad (5.11.10)$$

Here, $(PWI)^{(h)}_k$ and $(EPYTI)_k$ are defined in Eqs. (5.11.8c), and (5.11.9d).

5.14 Summary of time-volume-averaged interfacial balance equations

Time-volume-averaged interfacial mass, momentum, total energy, internal energy, and enthalpy balance equations are listed in this section.

5.14.1 Time-volume-averaged interfacial mass balance equation

$$\gamma_v {}^{3i}\langle \rho_k\rangle_{LF} \left(\frac{\partial \alpha_k}{\partial t} + {}^{3i}\langle \underline{U}_k\rangle_{LF} \cdot \nabla \alpha_k \right) + (MTI)_k$$

$$+ \gamma_v {}^{3i}\langle \rho_f\rangle_{LF} \left(\frac{\partial \alpha_f}{\partial t} + {}^{3i}\langle \underline{U}_f\rangle_{LF} \cdot \nabla \alpha_f \right) + (MTI)_f = 0$$

$$(5.4.1)$$

5.14.2 Time-volume-averaged interfacial linear momentum balance equation (including the capillary pressure difference)

$$\begin{aligned}
&\gamma_v\left({}^{3i}\langle P_k\rangle_{LF}\underline{\underline{I}} - {}^{3i}\langle\underline{\underline{\tau}}_k\rangle_{LF}\right)\cdot\nabla\alpha_k + (PTI)_k \\
&\quad - (VSTI)_k + \gamma_v\alpha_k{}^t\langle\Gamma_k\rangle{}^{3i}\langle\underline{U}_k\rangle_{LF} + (MMTI)_k \\
&\quad + \gamma_v\left({}^{3i}\langle P_f\rangle_{LF}\underline{\underline{I}} - {}^{3i}\langle\underline{\underline{\tau}}_f\rangle_{LF}\right)\cdot\nabla\alpha_f + (PTI)_f \\
&\quad - (VSTI)_f + \gamma_v\alpha_f{}^t\langle\Gamma_f\rangle{}^{3i}\langle\underline{U}_f\rangle_{LF} + (MMTI)_f \\
&= \gamma_v{}^{3i}\langle P_{ck}\rangle_{LF}\nabla\alpha_k - \gamma_v{}^{3i}\langle P_{cf}\rangle_{LF}\nabla\alpha_f \\
&\quad + (CPKI)_k - (CPFI)_f.
\end{aligned} \tag{5.6.1}$$

5.14.3 Time-volume-averaged interfacial total energy balance equation

$$\begin{aligned}
&\gamma_v\alpha_k\left({}^{t3i}\langle\dot{Q}_{kf}\rangle + {}^{t3i}\langle\dot{Q}_{wk}\rangle\right) + \gamma_v{}^{3i}\langle\underline{U}_k\rangle_{LF}{}^{3i}\langle P_k\rangle_{LF}\cdot\nabla\alpha_k \\
&- \gamma_v\left({}^{3i}\langle\underline{U}_k\rangle_{LF}\cdot{}^{3i}\langle\underline{\underline{\tau}}_k\rangle_{LF}\right)\cdot\nabla\alpha_k + (PWI)_k - (VWI)_k \\
&+ \gamma_v\alpha_k{}^t\langle\Gamma_k\rangle{}^{3i}\langle E_k\rangle_{LF} + (TETI)_k + \gamma_v\alpha_f\left({}^{t3i}\langle\dot{Q}_{fk}\rangle + {}^{t3i}\langle\dot{Q}_{wf}\rangle\right) \\
&+ \gamma_v{}^{3i}\langle\underline{U}_f\rangle_{LF}{}^{3i}\langle P_f\rangle_{LF}\cdot\nabla\alpha_f \\
&- \gamma_v\left({}^{3i}\langle\underline{U}_f\rangle_{LF}\cdot{}^{3i}\langle\underline{\underline{\tau}}_f\rangle_{LF}\right)\cdot\nabla\alpha_f + (PWI)_f \\
&- (VWI)_f + \gamma_v\alpha_f{}^t\langle\Gamma_f\rangle{}^{3i}\langle E_f\rangle_{LF} + (TETI)_f = 0
\end{aligned} \tag{5.8.1}$$

5.14.4 Time-volume-averaged interfacial internal energy balance equation

$$\begin{aligned}
&(PWI)_k^{(u)} + (VDI)_k + \gamma_v\alpha_k\left({}^{t3i}\langle\dot{Q}_{kf}\rangle + {}^{t3i}\langle\dot{Q}_{wk}\rangle\right) \\
&\quad + \gamma_v\alpha_k{}^t\langle\Gamma_k\rangle{}^{3i}\langle u_k\rangle_{LF} + (IETI)_k \\
&\quad + (PWI)_f^{(u)} + (VDI)_f + \gamma_v\alpha_f\left({}^{t3i}\langle\dot{Q}_{fk}\rangle + {}^{t3i}\langle\dot{Q}_{wf}\rangle\right) \\
&\quad + \gamma_v\alpha_f{}^t\langle\Gamma_f\rangle{}^{3i}\langle u_f\rangle_{LF} + (IETI)_f = 0
\end{aligned} \tag{5.10.1}$$

5.14.5 Time-volume-averaged interfacial enthalpy balance equation

$$
\begin{aligned}
&\gamma_v \alpha_k \left({}^{t3i}\langle \dot{Q}_{kf} \rangle + {}^{t3i}\langle \dot{Q}_{wk} \rangle \right) - \gamma_v\, {}^{3i}\langle P_k \rangle_{LF} \frac{d\alpha_k}{dt_k} + (PWI)_k^{(h)} \\
&+ (VDI)_k + \gamma_v \alpha_k\, {}^{t}\langle \Gamma_k \rangle^{3i} \langle h_k \rangle_{LF} + (EPYTI)_k \\
&+ \gamma_v \alpha_f \left({}^{t3i}\langle \dot{Q}_{fk} \rangle + {}^{t3i}\langle \dot{Q}_{wf} \rangle \right) - \gamma_v\, {}^{3i}\langle P_f \rangle_{LF} \frac{d\alpha_f}{dt_f} + (PWI)_f^{(h)} \\
&+ (VDI)_f + \gamma_v \alpha_f\, {}^{t}\langle \Gamma_f \rangle^{3i} \langle h_f \rangle_{LF} + (EPYTI)_f = 0 \qquad (5.12.1)
\end{aligned}
$$

6 Time averaging in relation to local volume averaging and time-volume averaging versus volume-time averaging

In this chapter, we discuss (1) time averaging in relation to local volume averaging and (2) proper order of time-volume averaging versus volume-time averaging.

6.1 Time averaging in relation to local volume averaging

The averaging procedure in multiphase mechanics must be related and can be understood by considering the basis of experimental observation [4]. The relative magnitudes of three quantities determine the method and meaning of averaging. They are the size of the dispersed phase, the spacing between the elements of the dispersed phase, and the volume observed. When applied to a two-phase boiling system, they become the size of bubbles, the mean spacing between bubbles, and the size of observation "window" (or any probe of finite size). For a one-dimensional system, these quantities are bubble size D, bubble spacing S, and observation window or slit width L.

The typical boiling flow system encountered in most engineering applications is the one with $D \sim S$ and $L < D$ as shown in Fig. 6.1a in one dimension, such as bubbles in liquid in a tube. In this case, strong interactions exist between phases 1 and 2. Spacewise, the bubble distribution over x is shown in Fig. 6.1b, where Δx_1, Δx_2, and Δx_3 are spaces occupied by bubbles at time t and the volume fraction of phase 1 is given by

$$\alpha_{1x} = (\Delta x_1 + \Delta x_2 + \Delta x_3)\,/X, \tag{6.1.1}$$

which is averaged over length X over a number of bubbles. If this flow system is observed at position x_0 over time T^*, then a record as shown in Fig. 6.1c will be obtained with t proportional to x because of constant velocity. In this case, the residence time fraction of phase 1 is given by

$$\alpha_{1t} = (\Delta t_1 + \Delta t_2 + \Delta t_3)\,/\,T^*. \tag{6.1.2}$$

Both Ishii [29] and Delhaye and Archard [24] specified that T^* must be chosen to be "large compared to turbulent fluctuation and small compared to overall flow fluctuations," that is, over time T^*, where velocity is nearly constant (in magnitude and direction). In such a special case, of course,

$$\alpha_{1t} = \alpha_{1x}, \tag{6.1.3}$$

and hence the volume fraction of phase 1 based on time average is in effect equal to local volume average over a volume represented by length X.

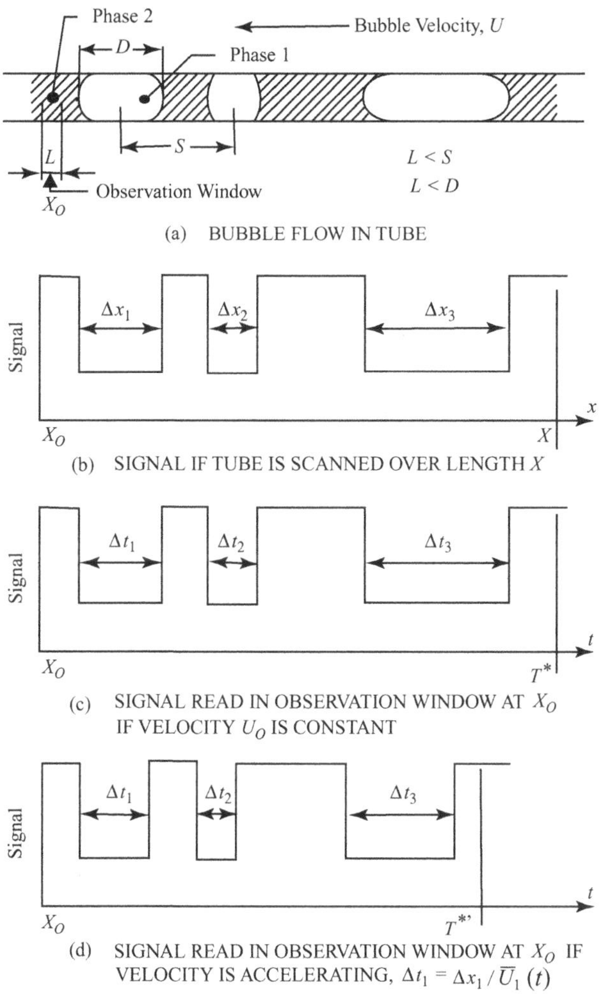

Fig. 6.1. Volume fraction in one-dimensional flow.

A different situation exists in the case of transient flow. This is illustrated in Fig. 6.1d for accelerating flow of speed $U(t)$ of all parts of the same system such that bubbles at X reach x_0 at time $T^{*\prime}$. In this case, the time-averaged volume fraction α_{1t} represents fractional residence time of phase 1 and is not equal to the volume-averaged volume fraction α_{1x}:

$$\alpha_{1t} = \left(\Delta t_1 + \Delta t_2 + \Delta t_3\right) / T^{*\prime} \neq \alpha_{1x}. \qquad (6.1.4)$$

Equivalence to α_{1x} is obtained with a time average weighted with velocity

$$\begin{aligned} \alpha'_{1t}\big|_{x_0} &= (\overline{U}_1 \Delta t_1 + \overline{U}_2 \Delta t_2 + \overline{U}_3 \Delta t_3) / \int_0^{T^{*\prime}} U dt, \qquad (6.1.5) \\ &\simeq \alpha_{1x}\big|_{x_0 \ to \ X(at \ t=0)} \end{aligned}$$

where $\overline{U}_1$ is the mean velocity over time interval Δt_1, and $\overline{U}_1 \Delta t_1 = \Delta x_1$, and so forth. In this way, the mean volume fraction of 1 can be identified in terms of time average. The dynamics of the system are determined by α'_{1t}, not α_{1t}. Moreover,

$$X = \int_0^{T^{*\prime}} U dt = \overline{U} T^{*\prime}. \qquad (6.1.6)$$

Because U, $\overline{U}$, $\overline{U}_i$ $(i = 1,2,3,\ldots)$, are all functions of time at x_0, or $U = U\left(x_0, t\right)$, there is no conceptual difficulty in carrying out velocity-weighted, time-averaging procedures with

$$\begin{aligned} \alpha'_{1t}\big|_{x_0} &= ((\overline{U}_1/\overline{U})\Delta t_1 + (\overline{U}_2/\overline{U})\Delta t_2 + (\overline{U}_3/\overline{U})\Delta t_3)/T^{*\prime} \\ &\equiv T_1^*/T^{*\prime} \simeq \alpha_{1x}\big|_{x_0 \ to \ X}, \qquad (6.1.7) \end{aligned}$$

which defines T_1^*, and with improved accuracy for small $T^{*\prime}$. The basis for the averaging is for

$$(X/\overline{U}) > T^{*\prime} > (L/\overline{U}).$$

In summary, the time-averaged volume fraction of phase 1, α_{1t}, is not, in general, equal to local volume-averaged volume fraction α_{1x}. $\alpha_{1x} \equiv \alpha_{1t}$, if and only if the velocity field is one-dimensional and uniform. Furthermore, local volume averaging from the beginning gives $\alpha_{1x}(t)$; subsequent time averaging still gives a time-averaged volume fraction. Only the volume fraction of the phase provides for the cumulative nature of thermodynamic properties. Therefore, volume averaging must precede time averaging to give physically meaningful quantities.

6.2 Time-volume averaging versus volume-time averaging

The significance of first performing volume averaging of the phasic conservation equations and their associated interfacial balance equations, followed by time averaging, has been pointed out previously. This order of averaging preserves the distinction of the dynamic phases in a multiphase system, such as droplets or bubbles of different sizes, or particles of the same size and materials, but of different electric charges. Eulerian time averaging from the beginning will remove this distinction unless suitable conditional averaging is used. Simple time averaging leads to the fractional residence time of a phase rather than the volume fraction of a phase. This fractional residence time* of a phase becomes equal to the

* Fractional residence time was referred to as local time fraction or time-averaged phase density function in Ref. [29].

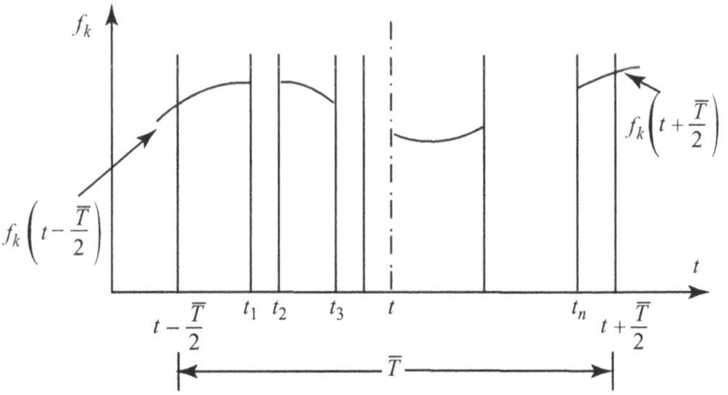

Fig. 6.2. Variation of f_k with time at fixed point.

physical volume fraction only in the case of one-dimensional, uniform motion of incompressible flows.

The foregoing discussion clearly contradicts the conclusion reached by Delhaye and Archard [24], who stated that the order of time-volume averaging is interchangeable and gave a mathematical proof to support their claim. Unfortunately, their proof was in error because of improper application of the Leibnitz rule for the differentiation of an integral [4]. A brief recapitulation of their derivation follows.

Consider the variation of any property f_k associated with phase k, such as density, temperature, or velocity, as seen by an observer at a fixed point in multiphase flows. Because phase k passes through the point intermittently, f_k would have the appearance shown in Fig. 6.2 [24,29].

Delhaye and Archard [24] considered the time interval $(t - \frac{\overline{T}}{2}, t + \frac{\overline{T}}{2})$ centered at the instant t and denoted the cumulated residence time of phase k in the interval by $[T_k]$. The averaging time interval $\overline{T}$ was taken to be a constant.

Referring to Fig. 6.2, we may write

$$\int_{[T_k]} f_k dt = \int_{\eta=t-\overline{T}/2}^{\eta=t_1} f_k(\eta)d\eta + \int_{\eta=t_2}^{\eta=t_3} f_k(\eta) \qquad (6.2.1)$$
$$+ \cdots + \int_{\eta=t_\eta}^{\eta=t+\overline{T}/2} f_k(\eta)d\eta,$$

where η is the dummy variable of integration. The Leibnitz rule for differentiation of an integral was applied, and at the same time, set

$$\frac{dt_i}{dt} = 0, \quad for\, i = 1, 2, \ldots n. \qquad (6.2.2)$$

The result was

$$\frac{\partial}{\partial t}\int_{[T_k]} f_k\, dt = f_k\left(t + \frac{\overline{T}}{2}\right) - f_k\left(t - \frac{\overline{T}}{2}\right). \qquad (6.2.3)$$

That is the time derivative of the integral defined in Eq. (6.2.1) depends only on the values of the integral f_k evaluated at the two end points.

Let us pause and inquire into the physical meaning of $\frac{\partial}{\partial t}\int_{[T_k]} f_k\, dt$ under the condition defined in Eq. (6.2.2). On the one hand, we have from elementary calculus

$$\frac{\partial}{\partial t}\int_{[T_k]} f_k\, dt = \lim_{\Delta t \to 0} \frac{\left(\int_{[T_k]} f_k\, dt\right)_{t+\Delta t} - \left(\int_{[T_k]} f_k\, dt\right)_t}{\Delta t}. \qquad (6.2.4)$$

On the other hand, we note that for time averaging to be physically meaningful in multiphase flows, the averaging duration $\overline{T}$ must encompass a sufficiently large number of interfaces (i.e., it must be large relative to the inverse of the passage frequency of the phase interface υ_s). At the same time, it must be small compared with that required for the mixture flowing at a representative velocity U through the characteristic dimension L of the system. Hence,

$$(L/U) \gg \overline{T} \gg (1/v_s). \tag{6.2.5}$$

In other words, Δt in Eq. (6.2.4) should never be allowed to approach zero. It must be finite. Now for time t, phase k will first leave the observation point at time $t_1 - (t - \overline{T}/2)$ after the initial arrival of phase k. For time $t + \Delta t$, phase k will, in general, *not* leave the observation point at the same time subsequent to its initial arrival. The same can be said for the "arrival" time t_2 and "departure" time t_3, and so on. Hence, t_i is not independent of t, and the use of Eq. (6.2.2) is not physically realizable in multiphase flows. Consequently, Eqs. (6.2.2) and (6.2.3) are invalid; so is the conclusion reached by [24].

An additional difficulty is associated with the application of Eulerian time averaging to the phasic conservation equation from the beginning. The difficulty stems from the fact that the time interval $\overline{T}$ chosen for averaging is not intrinsic to the structure of the multiphase medium under consideration, but instead depends strongly on the convection velocity. In most engineering systems, wide ranges of velocities often exist; hence, they may not be characterized by a single time scale. This is in contrast to the length scale associated with local volume averaging, which is independent of the flow velocity.

7 Novel porous media formulation for single phase and single phase with multicomponent applications

A generic three-dimensional, time-dependent family of COMMIX codes [37] for single phase with multicomponent has been developed based on the novel porous media formulation. In the novel porous media formulation, each computational cell is characterized by volume porosity (γ_v), directional surface porosities (γ_{Ax}, γ_{Ay}, and γ_{Az} for Cartesian coordinates), distributed resistance (${}^{3i}\langle \underline{R}_k \rangle$), and distributed heat source and sink (${}^{3i}\langle S_k \rangle$) through input data. The formulation can readily be brought back to the conventional porous media formulation and continuum formulation by setting $\gamma_{Ax} = \gamma_{Ay} = \gamma_{Az} = 1$ and $\gamma_v = \gamma_{Ax} = \gamma_{Ay} = \gamma_{Az} = 1$, respectively [3,37]. In fact, in many practical engineering analyses, the computational domain can be subdivided into many subdomains. Some of these subdomains are more suitable to use the conventional porous media formulation (e.g., reactor core) and others are appropriate to use the continuum formulation (e.g., for a reactor upper plenum, where there are no solid structures). With appropriate input of the volume porosity, directional surface porosities, distributed resistance, and distributed heat source and sink for each computation cell, the novel porous media formulation enables engineers and scientists to solve challenging real world engineering problems with complex stationary solid structures.

Most of our experience with the novel porous media formulation had been in single phase and single phase with multicomponent applications, and limited in two-phase flow with conservation equations approximated as a set of partial differential equations. Recently, we developed the novel porous media formulation from single phase to multiphase as described in this book (see Chapter 5). These conservation equations are in differential-integral form and are not a set of partial differential equations, as currently appear in most literature on two-phase and multiphase flow. To the best of our knowledge, no one has ever solved the conservation equations in the differential-integral forms that are presented in this book. Engineers and scientists have been working on two-phase flow throughout the world since the Industrial Revolution in late 18th century. It is startling to realize that the correct conservation equations of mass, momentum, and energy for two-phase flow analysis have not been used since the invention of steam engines and boilers more than 100 years ago.

The contributions from the novel porous media formulation for single phase and single phase with multicomponent [37,40] are exemplified by numerous studies: (1) Von Karmann vortex shedding analysis [38,40–42]; (2) shear-driven cavity flow analysis [39–42]; (3) pressurized thermal shock

(PTS) analysis to help Electric Power Research Institute (EPRI) resolving the PTS issue [43,44]; (4) heat exchanger analysis to help industry improving their product and performance [45,46]; (5) in-vessel analysis, including all components inside a reactor vessel, to understand coupling effects between core, upper plenum, downcomer, lower plenum, and other reactor components [47,48]; (6) analysis of natural convection phenomena in a prototypic pressurized water reactor (PWR) during a postulated degraded core accident to delineate natural convection flow patterns and to show the importance of a steam generator's heat capacity effect on accident progression and consequence [49,50]; and (7) analyses of large-scale tests for the AP-600 passive containment cooling system (PCCS) in order to improve its design and performance [51,52] and the PCCS used the condensation and evaporation to remove heat generated during a postulated design-based reactor accident (DBA). It is well known that heat transfer via phase change is one of the most efficient means. The results from the COMMIX code of these analyses cited previously were compared with corresponding experimental data, and the agreement is good.

Four cases – (1), (2), (6), and (7) – have been selected for detailed presentation, including experimental setup and comparison of calculated results from the COMMIX code with novel porous media formulation for single phase and corresponding experimental data. Cases (1) and (2) show that the COMMIX code is not only correctly implemented, but also capable of computing detailed microflow fields with fine computational mesh and high-order differencing scheme. Cases

(6) and (7) are grouped to demonstrate that the COMMIX code can capture essential flow fields and temperature distributions with coarse computational mesh and local volume-averaged dependent variables for solving practical engineering problems. The following are detailed presentations of each case.

7.1 COMMIX code capable of computing detailed microflow fields with fine computational mesh and high-order differencing scheme

We performed Von Karmann vortex shedding analysis by using the COMMIX code with the LECUSSO (Local Exact Consistent Upwind Scheme of Second Order) high-order differencing scheme [40] and compared the results with experimental data of Davis et al. [38] shown in Section 7.1.1. Numerical calculations were carried out for two-dimensional, shear-driven cavity flows by using the COMMIX code with both the LECUSSO and the QUICK (Quadratic Upstream Interpolation for Convective Kinematics) [41] high-order differencing schemes and with the FRAM [42], a nonlinear damping algorithm shown in Section 7.1.2.

7.1.1 Case (1): Von Karmann vortex shedding analysis

The Von Karmann vortex shedding analysis study is a test problem of momentum transport [38,40]. Figure 7.1 shows the configuration of the experiment by Davis et al. [38], and the computational modeling is listed here for Case (1a):

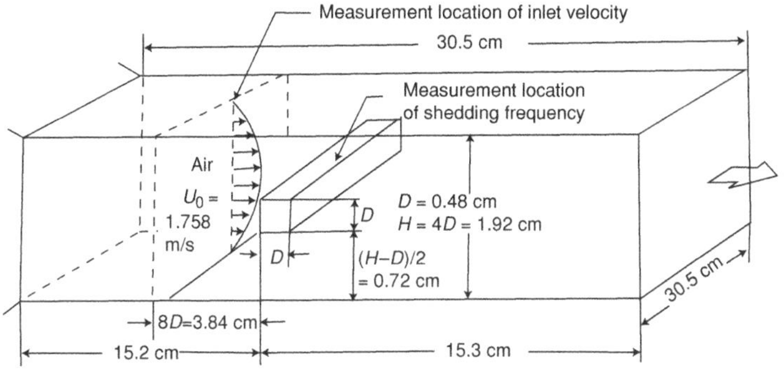

(a) Configuration of experiment (NBS Low Velocity Airflow Facility)
$Re = U_0D/v = 550$ (at 20°C)

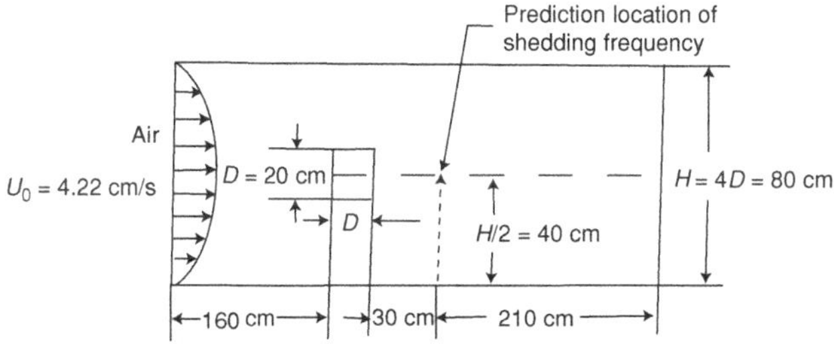

(b) 2-D model of computation $Re = U_0D/v = 550$ (at 20°C)

Fig. 7.1. Configuration of experimental facility and computational modeling for Von Karmann vortex shedding analysis.

Case (1a)

Fluid: air

Square obstacle: length $D = 0.2$ m

Flow channel: confined channel; channel width $H = 4 \times D = 0.8$ m

Inlet velocity distribution: measurement with maximum velocity $U_0 = 0.0422$ m/s, and $Re = U_0 \times D/\upsilon = 0.0422 \times 0.2/1.536E-5 = 550$ (at 20°C)

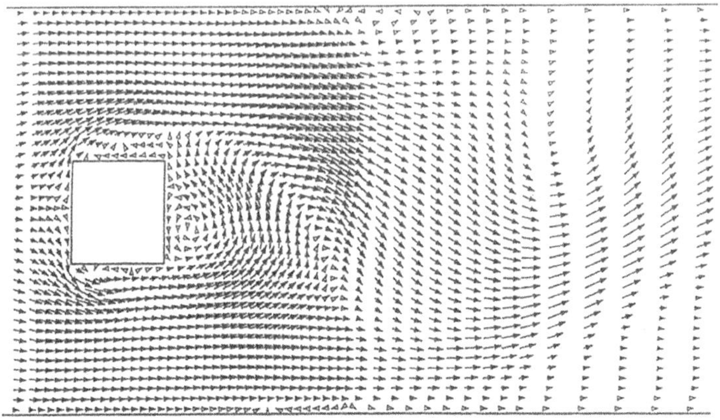

$J = 1$

0.30 m/s

Time: 151.90 s

LECUSSO Scheme

$Re = 550$ (Air Fluid)

Fig. 7.2. Velocity vector plot from Von Karmann vortex shedding analysis.

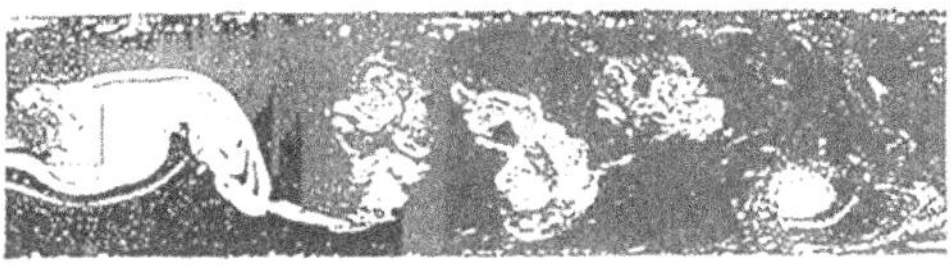

(a) Experiment (streakline)

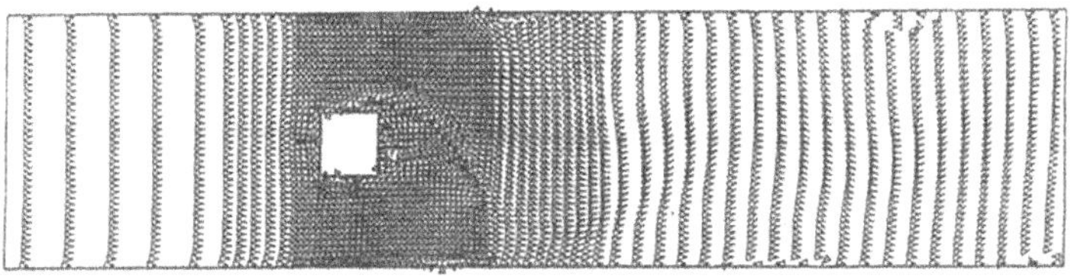

$J = 1$

1.00 m/s

Time: 151.90 s

(b) Prediction (velocity vector)

LECUSSO Scheme

$Re = 550$ (Air Fluid)

Fig. 7.3. Experimentally determined streakline and predicted velocity vector.

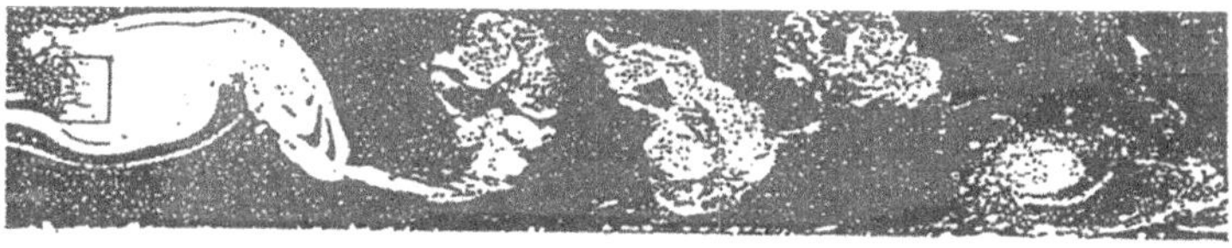

(a) Experiment

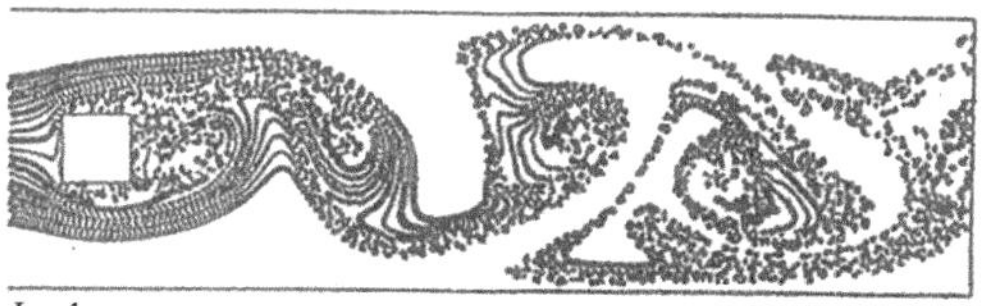

$J = 1$
Time: 125.00 s

LECUSSO Scheme
$Re = 550$ (Air Fluid)

(b) Prediction

Fig. 7.4. Experimentally determined and predicted streaklines from Von Karmann vortex shedding analysis.

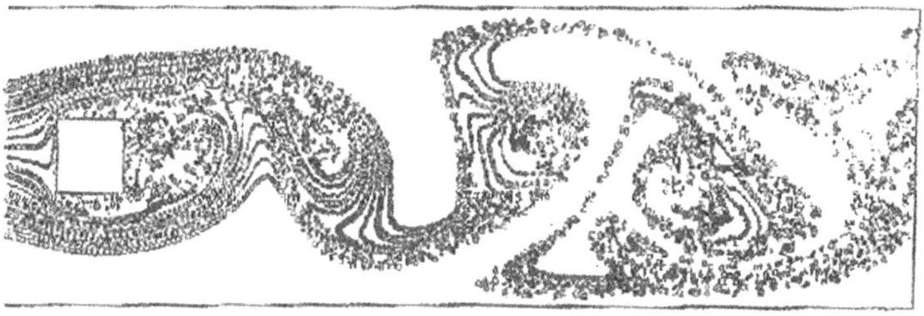

J = 1
Time: 125.00

LECUSSO Scheme
$Re = 550$ (Air Fluid)

Fig. 7.5. Enlarged view of computed streaklines shown in Fig. 7.4.

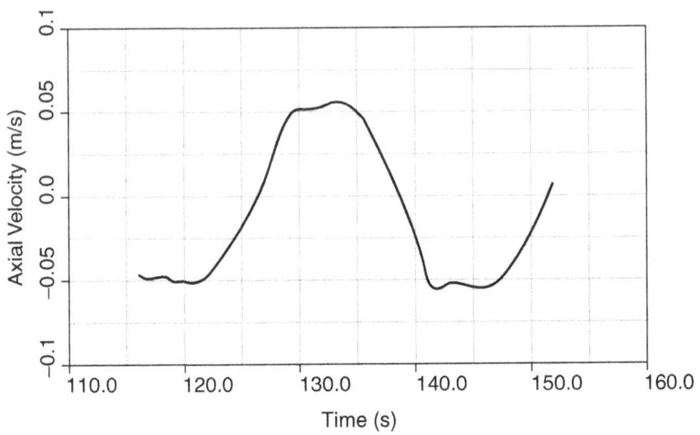

Fig. 7.6. Time variation of fluctuating axial velocity in wake region at $Re = 550$ during Von Karmann vortex shedding analysis.

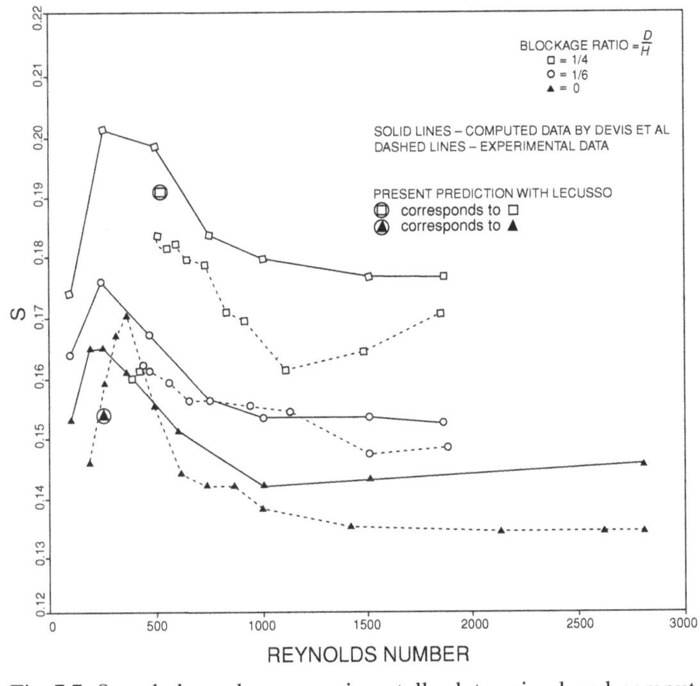

Fig. 7.7. Strouhal numbers experimentally determined and computed by Davis et al. [38] and predicted by current LECUSSO [40].

The results from the COMMIX code are listed here:
Prediction of Strouhal number S:

$$S = F \times D/U_0 = 0.04032 \times 0.2/0.0422 = 0.191$$
$$F = \text{frequency} = 1/T = 1/24.8 = 0.04032$$
$$T = \text{period} = (151.3 - 126.5)\ \text{s} = 24.8\ \text{s (from Fig. 7.6)}$$

Measurement of Strouhal number $S = 0.182$ (from Fig. 7.7).

The computational modeling is listed below for Case (1b):

Case 1b

Fluid: air

Square obstacle: length $D = 0.02$ m

Flow channel: open channel; channel width $H = 6 \times D = 0.12$ m

Inlet velocity distribution: measurement with maximum velocity $U_0 = 0.0127$ m/s, and $Re = U_0 \times D/\upsilon = 0.0127 \times 0.02/1.011E-6 = 251$ (at 20°C)

The results from the COMMIX code are listed here:
Prediction of Strouhal number S:

$$S = F \times D/U_0 = 0.0980 \times 0.02/0.0127 = 0.154$$
$$F = \text{frequency} = 1/T = 1/10.2 = 0.0980$$
$$T = \text{period} = (114.5 - 104.3)\ \text{s} = 10.2\ \text{s (from Fig. 7.8)}$$

Measurement of Strouhal number $S = 0.159$ (from Fig. 7.7).

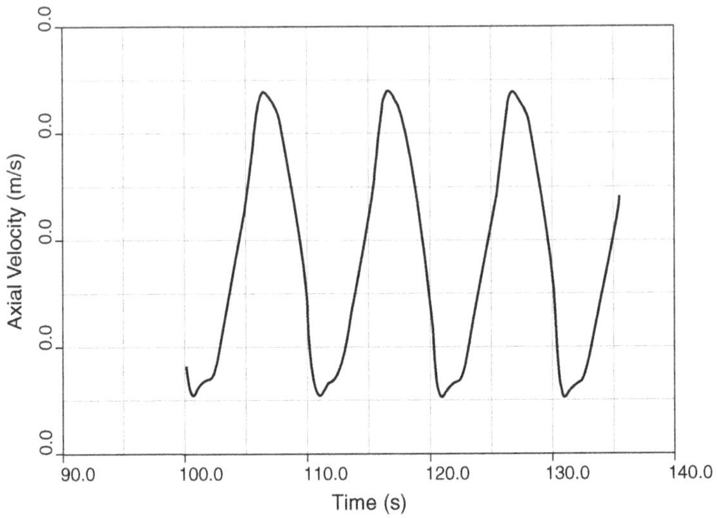

Fig. 7.8. Time variation of fluctuating axial velocity in wake region at $Re = 251$ during Von Karmann vortex shedding analysis.

In summary, comparisons of the streaklines and Strouhal numbers calculated by the COMMIX code with the high-order differencing LECUSSO scheme [40] and corresponding experimental data [38], as shown in Figs. 7.4 and 7.7, respectively, indicate good agreement. This is a clear indication that the COMMIX code is properly programmed and capable of computing the microflow field. We noted that the Von Karmann vortex shedding problem is considered as a classic example for validating a code's accuracy and capability.

7.1.2 Case (2): Shear-driven cavity flow analysis

Numerical calculations were performed for two-dimensional, shear-driven cavity flows by using the COMMIX code with

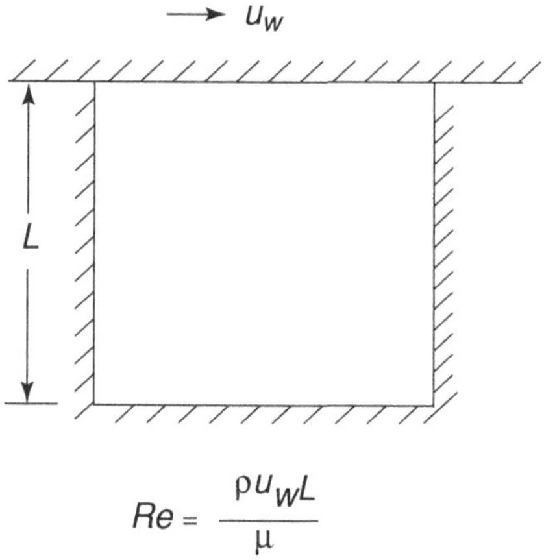

Fig. 7.9. Computational region for shear-driven cavity flows.

both LECUSSO and QUICK [41] high-order differencing schemes and FRAN [42] with $\Delta x = \Delta z = 0.0125$ m, incompressible fluid at temperature 20°C, as shown in Fig. 7.9.

Two-dimensional, shear-driven cavity flow represents an excellent test for evaluating convective differencing schemes because of the large streamline-to-grid skewness present over most of the flow region and the existence of several relatively large recirculation regions. The calculation results were compared with those of Ghia et al. [39], who used grids that were quite fine, such as 257 × 257 meshes, and whose results can be considered almost exact solutions.

The results of Case (2) from the COMMIX code are listed here:

In summary, a comparison of the normalized lateral velocity component obtained from the COMMIX code with high-order differencing schemes QUICK and LECUSSO [40] (40 × 40 meshes) as shown in Fig. 7.10 with the calculations of Ghia et al. [39] at $Re = 1000$ for shear-driven

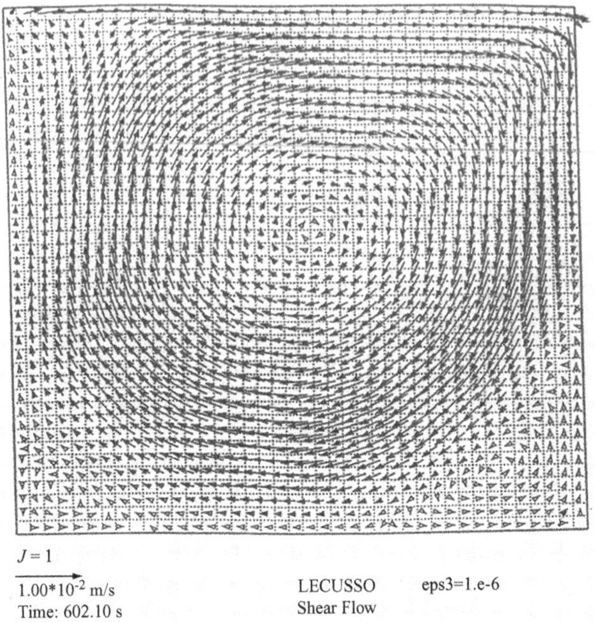

Fig. 7.10. Velocity vector at Re = 1,000 (40 × 40 meshes) for shear-driven cavity flows.

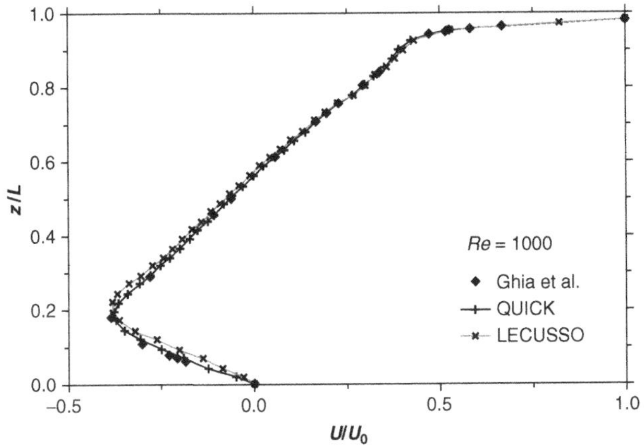

Fig. 7.11. Lateral velocity component from present calculations and from calculation of Ghia et al. [39] at Re = 1,000 (40 × 40 meshes) for shear-driven cavity flows.

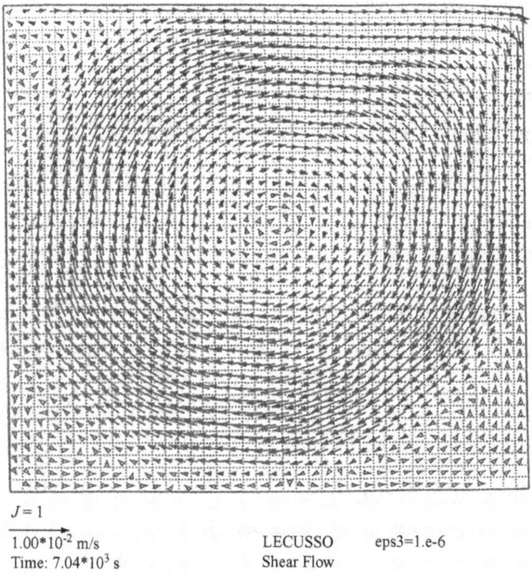

Fig. 7.12. Velocity vector at Re = 5,000 (40 × 40 meshes) for shear-driven cavity flows.

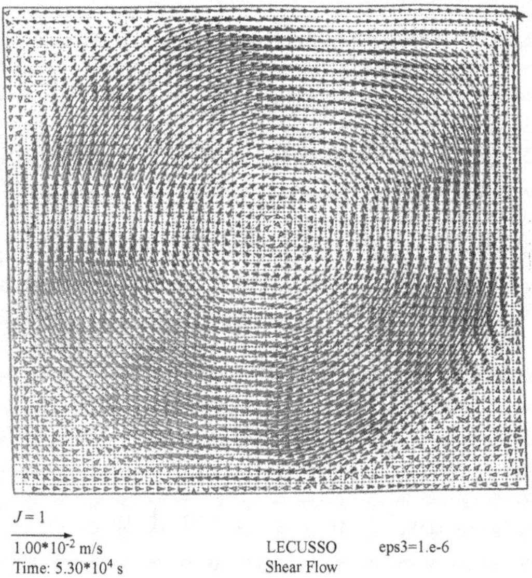

Fig. 7.13. Velocity vector at Re = 10,000 for shear-driven cavity flows.

cavity flows presented in Fig. 7.11, is in excellent agreement. We note that the COMMIX code with the LECUSSO scheme gives a stable solution of such high Reynolds numbers as shown in Figs. 7.12 and 7.13, but that the QUICK scheme failed. In addition, a comparison of corner vortices between the velocity vector predicted from the COMMIX code with the LECUSSO scheme at $Re = 1000$, 5000, and 10,000, as shown in Figs. 7.11–7.13, respectively, with the calculated results from Ghia et al. [39], as shown in Fig. 7.14, are in good agreement. This shear-driven cavity flow is another classic test for code accuracy and capability. We pointed out that Ref. [40] has a few typical classic test problems involving temperature distributions that prove the accuracy and capability of the COMMIX code beyond any doubt.

7.1.3 Some observations about higher-order differencing schemes

To carry out analyses such as Von Karmann vortex shedding analyses [Case (1)] and shear-driven cavity flows [Case (2)], one must implement the high-order differencing schemes LECUSSO [40], QUICK [41], and FRAM [42] into the COMMIX code. Based on the numerical results of the test problems, the following observations are obtained:

1. The high-order differencing schemes, LECUSSO and QUICK, with an oscillation dumping technique such as FRAM [42], agree well with experimental data. However, the agreement becomes poor without using the oscillation dumping technique.

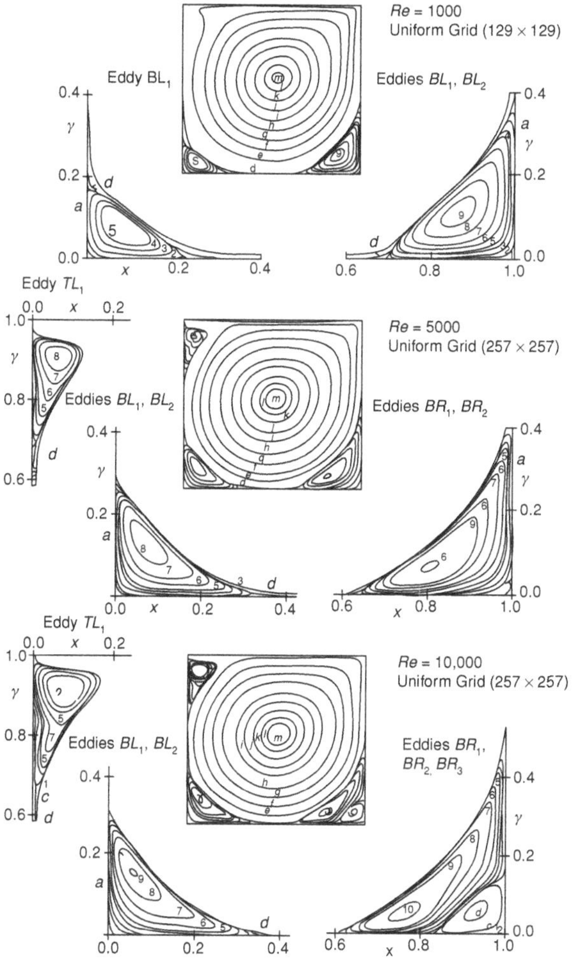

Fig. 7.14. Streamline pattern for primary, secondary, and additional corner vortices obtained by Ghia et al. [39].

2. From the point of view of stability, for a problem with uniform mesh, both high-order schemes LECUSSO [40] and QUICK [41] are stable. LECUSSO appears to be more stable than QUICK, but QUICK has a smaller truncation error. Both schemes need improvements in dealing with nonuniform meshes.

7.2 COMMIX code capable of capturing essential both macroflow field and macrotemperature distribution with a coarse computational mesh

We have successfully carried out analyses of both the natural convection phenomena in a prototypic PWR during a postulated degraded core accident [49,50] and large-scale tests for the AP-600 PCCS [51,52]. We have also conducted comparisons between the calculated results obtained from the COMMIX code with the novel porous media formulation and corresponding experimental data; the results are presented, respectively, in Sections 7.2.1 and 7.2.2.

7.2.1 Case (6): Natural convection phenomena in a prototypical pressurized water reactor during a postulated degraded core accident

The objective of the analysis of natural convection phenomena in a prototypic PWR during a postulated degraded core accident [50,51] is twofold: (1) to delineate natural convection patterns and their impact on the accident's progression and consequences, and (2) to validate the COMMIX code.

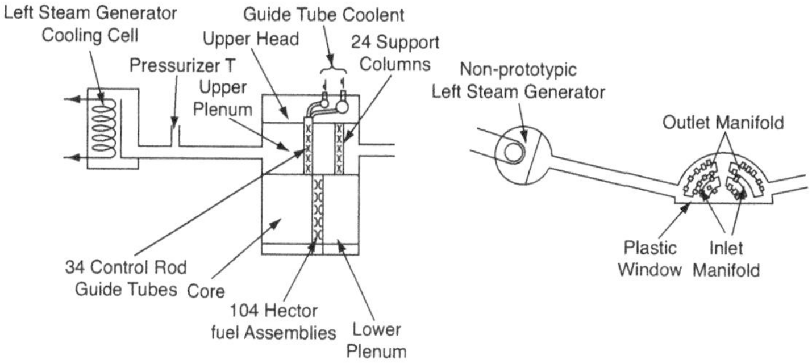

Fig. 7.15. Schematic diagrams for Westinghouse natural convection test.

Figure 7.15 presents schematic diagrams for Westinghouse's natural convection test. The experimental apparatus is a typical Westinghouse's four-loop PWR at 1/7 scale with symmetry at the middle of reactor vessel. The detailed mockup of the actual system in the experimental apparatus is also shown in Fig. 7.15. Figure 7.16 shows schematic layout of the three-dimensional model of the Westinghouse

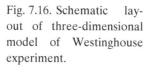

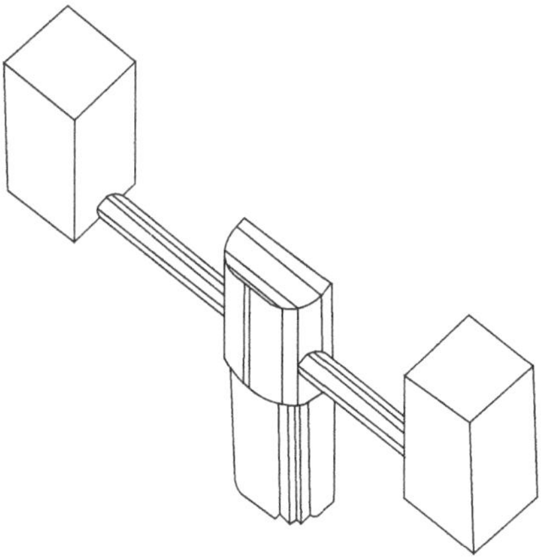

Fig. 7.16. Schematic layout of three-dimensional model of Westinghouse experiment.

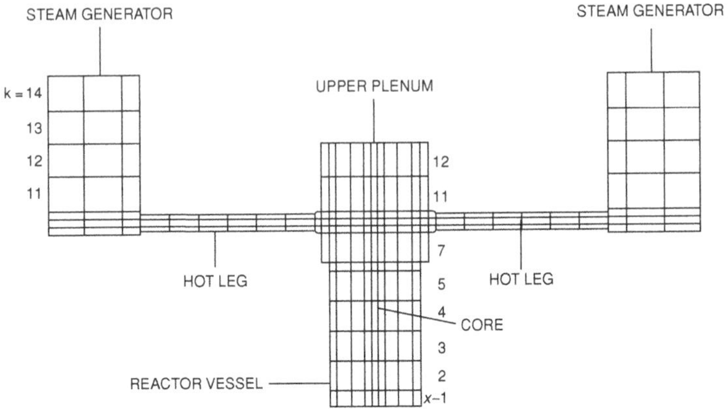

Fig. 7.17. Axial partitioning used in numerical model ($I = 8$).

experiment and axial partitioning used in the numerical model ($I = 8$), as presented in Fig. 7.17. In the original design of Westinghouse's experiment, both steam generators on the right and left sides of the reactor vessel are identical and nonprototypical (not a tubular steam generator), as shown in Figs. 7.18a, 7.19, and 7.20. At the later stage of this experiment, Westinghouse decided to replace the right steam generator by a prototypical design of a tubular steam generator as shown in Fig. 7.18b. Accordingly, the numerical model for the right side of steam generator has been modified as shown in Fig. 7.18b. Figure 7.19 presents the horizontal partitioning used in the numerical model ($K = 9$), and Fig. 7.20 shows that the horizontal partition of the secondary cooling system ($K = 12$).

The calculated results from the COMMIX code presented here are corresponding to the experiment dated February 16, 1985, with low-pressure water and a tubular steam generator on the right side of the reactor vessel [50].

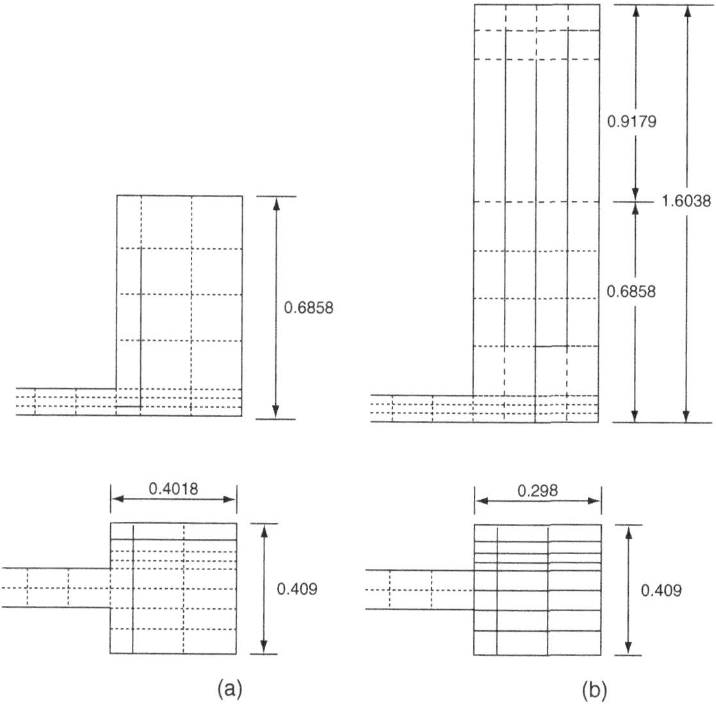

Fig. 7.18. Numerical models of right-side steam generator: (a) previous model and (b) new model of tubular steam generator.

Following are comparisons of numerical results from the COMMIX code with the novel porous media formulation for Case (6) [50], with corresponding Westinghouse experimental data [51].

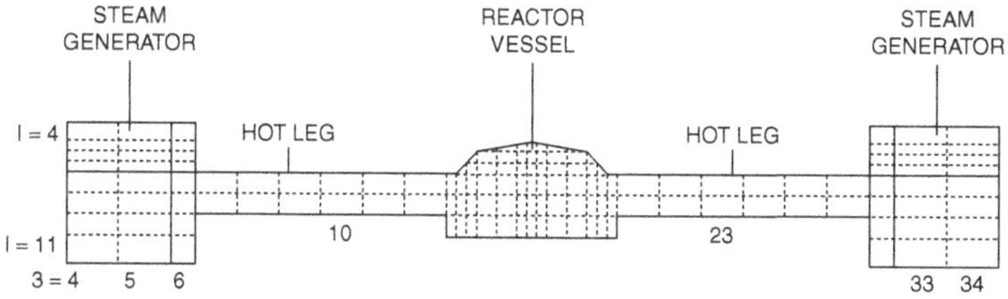

Fig. 7.19. Horizontal partitioning used in numerical model ($K = 9$).

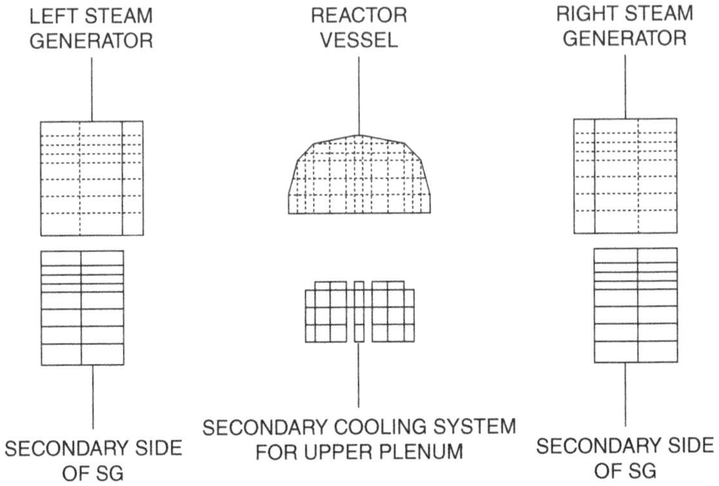

Fig. 7.20. Horizontal partitioning of secondary cooling system ($K = 12$).

7.2.1.1 Heat transfer

Table 7.1 shows a comparison of heat transfer between computed and measured values. The computed heat transfers are in very good agreement with the measured values.

Table 7.1. *Comparison of simulated and measured heat transfer*

Component	Computed Heat Transfer (kW)	Measured Heat Transfer (kW)
Upper internals	15.08	15.14
Upper plenum wall	4.65	4.67
Right steam generator	2.49	2.43
Left steam generator	5.60	5.66
Total	27.82	27.90
Core heating	28.70	28.70

Table 7.1 presents a comparison of simulated and measured heat transfer in various parts of the PWR system [50].

7.2.1.2 Natural convection patterns

Figures 7.21 and 7.22 present velocity field in the plane at $I = 8$ and 9, respectively, which clearly exhibits the following four large natural convection loops:

1. From lower plenum to core, to upper plenum, and then back down to lower plenum
2. From reactor vessel to left steam generator and then back to reactor vessel
3. From reactor vessel to right steam generator and then back to reactor vessel
4. Countercurrent fluid flow in hot legs and highly stratified

The preceding calculated natural convection patterns and stratification in hot legs were observed in the Westinghouse experiment and are in good agreement [50,51].

7.2.1.3 Temperature distribution

We calculated the temperature distribution from the COMMIX code, as shown in Fig. 7.23 at plane $I = 8$ and Fig. 7.24 at plane $I = 9$, which was compared with the measured temperature, and we observe the following:

1. *Lower Plenum*: The temperatures in the lower plenum are in very close agreement, at about 33°C to 34°C.

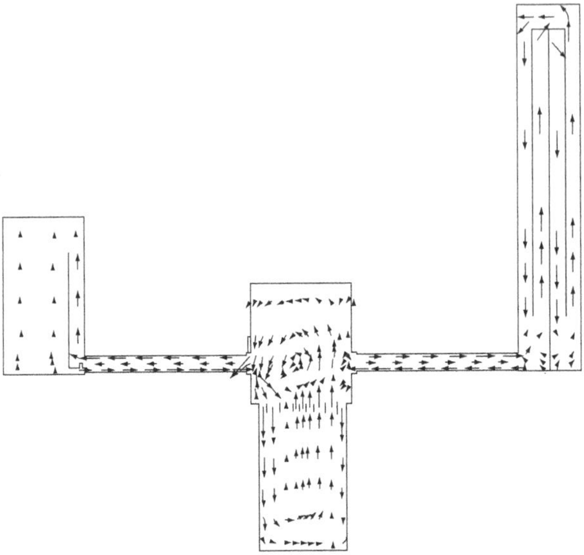

Fig. 7.21. Velocity field in plane ($I = 8$).

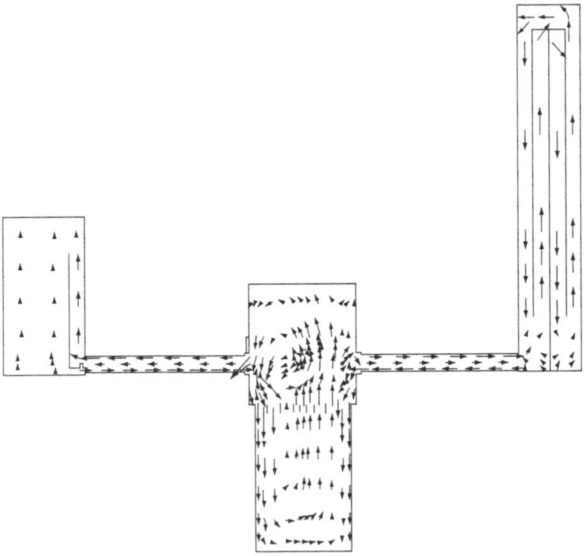

Fig. 7.22. Velocity field in plane ($I = 9$).

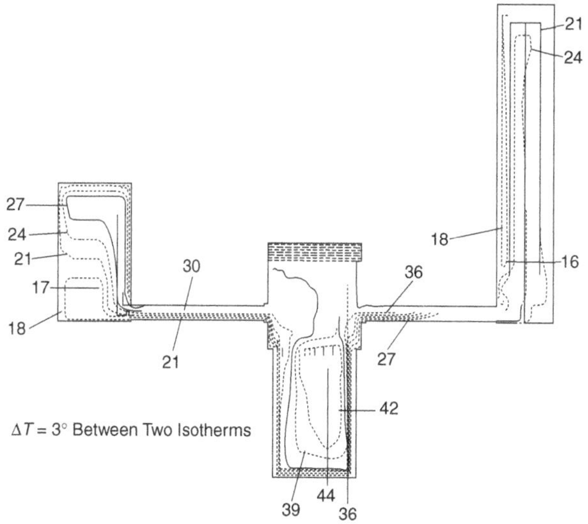

Fig. 7.23. Isotherms showing temparature distribution in plane ($I = 8$), °C.

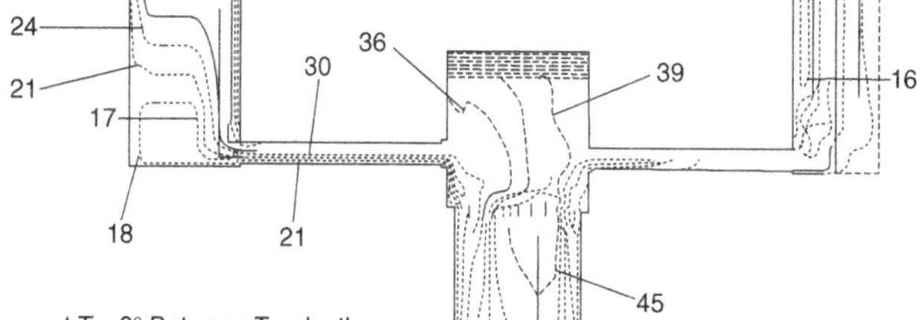

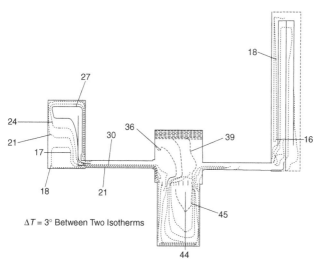

Fig. 7.24. Isotherms showing temparature distribution in plane ($I = 9$), °C.

2. *Core*: The temperatures in the core are also in good agreement. The left-side downcoming temperatures are about 30°C to 31°C. The right-side downcoming temperatures are about 36°C to 38°C. The central upper core temperatures are about 44°C to 46°C.
3. *Upper Head*: The temperatures in the upper head are in good agreement, in the range of 36°C to 40°C.
4. *Upper Plenum*: In the upper plenum, the range of computed temperatures, approximately 30°C to 41°C, is about the same as the measured range of temperatures from 30°C to 43°C.

In summary, numerical results obtained from the COMMIX code for the analysis of natural convection phenomena in a prototypic PWR during a postulated degraded accident are in good agreement with the corresponding experimental data. For the first time, the countercurrent stratified flow in a hot leg due to natural convection was calculated, which resulted from the heat capacity effect of the steam generator structures as a heat sink. This phenomenon was observed in the experimental date. We note that this is the first complete three-dimensional numerical simulation for a four-loop PWR ever done. Through this analysis, we have gained significant insights into natural convection during a postulated degraded core accident such as TMI-2 (Three Mile Island). With proper design of PWRs by introducing inherent and passive safety features, we can prevent TMI-2–type accidents and ensure the public safety of nuclear power plants.

7.2.2 Case (7): Analysis of large-scale tests for AP-600 passive containment cooling system

The objective of analysis of large-scale tests for the AP-600 PCCS [51,52] were twofold: (1) to validate the COMMIX code, and (2) to use the code once was validated for improving and optimizing the PCCS's design and performance at normal and abnormal operating conditions.

Figure 7.25 describes the AP-600 PCCS, and the numerical modeling for the PCCS is described in Figs. 7.26 and 7.27.

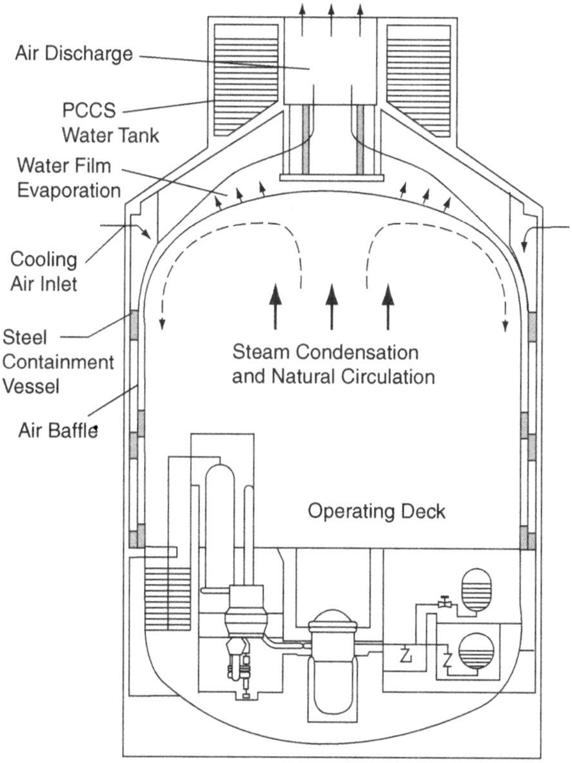

Fig. 7.25. AP-600 passive containment cooling system (PCCS).

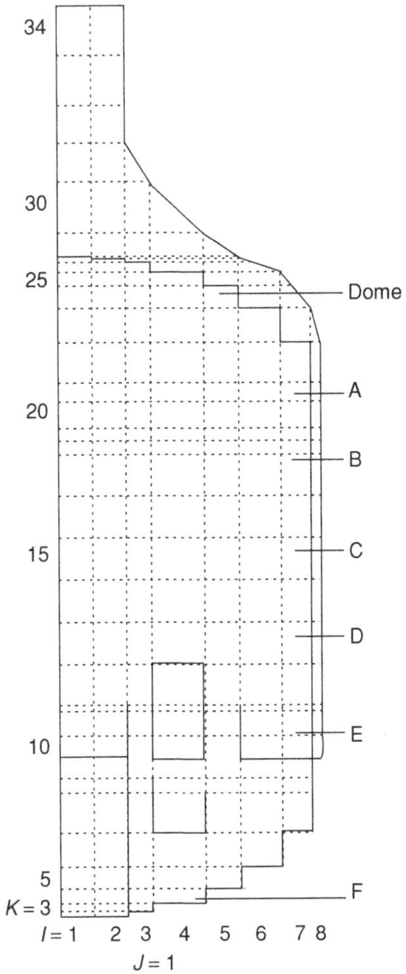

Fig. 7.26. Computational mesh representation for LST on *R*-*Z* cross section.

Figure 7.28 presents calculated results from COMMIX code for velocity, normalized temperature, and steam mass distributions at $J = 4$ of text 1. The comparison of the results from the COMMIX code and corresponding experimental measured data are shown in Figs. 7.29–7.38.

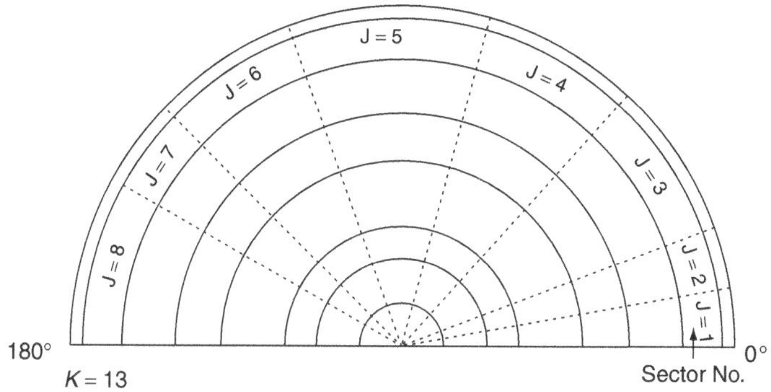

Fig. 7.27. Computational mesh representation for LST on R-θ cross section with $J_{max} = 8$.

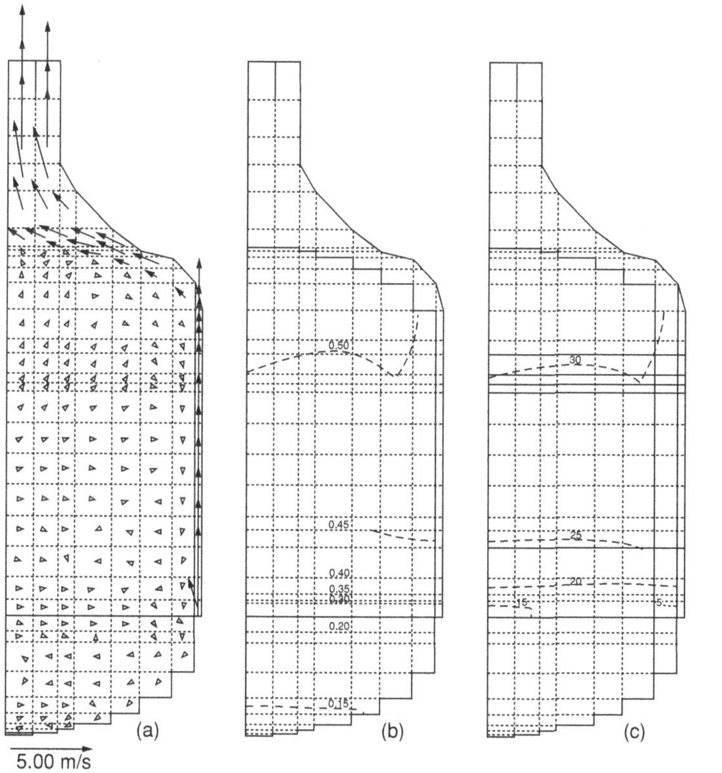

Fig. 7.28. (a) Velocity distribution at $J = 4$ of test 1, (b) normalized temperature distribution at $J = 4$ of test 1, and (c) steam mass fraction (%) distribution at $J = 4$ of test 1.

7.2.2.1 Circumferential temperature distribution

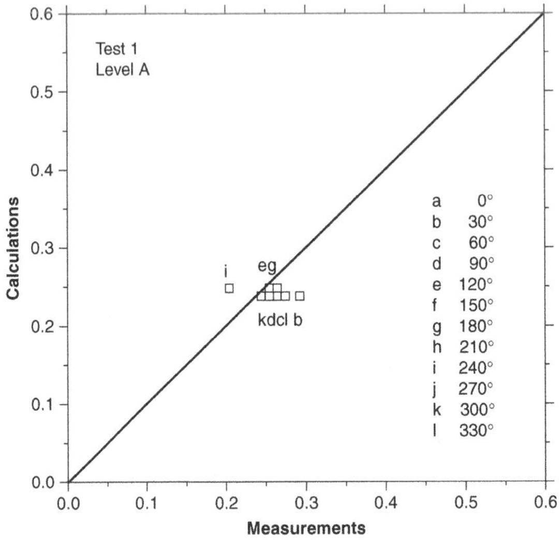

Fig. 7.29. Calculated and measured normalized circumferential wall temperature distribution of inside surface at top of cylindrical vessel of test 1.

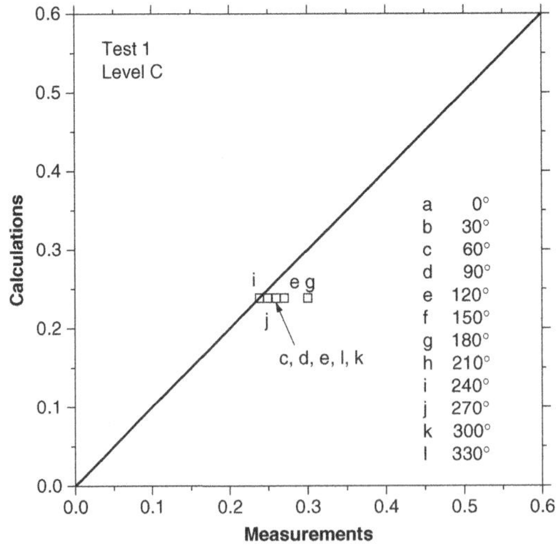

Fig. 7.30. Calculated and measured normalized circumferential wall temperature distribution of inside surface at middle of cylindrical vessel of test 1.

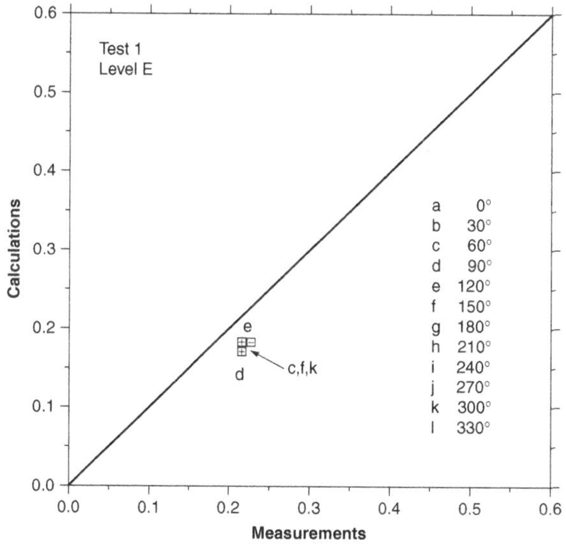

Fig. 7.31. Calculated and measured normalized circumferential wall temperature distribution of inside surface at bottom of cylindrical vessel of test 1.

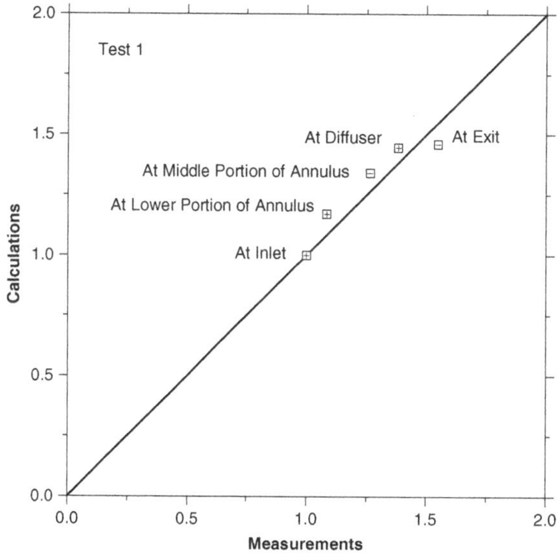

Fig. 7.32. Calculated and measured normalized temperature distribution of moist air in annulus at various elevations of test 1.

7.2.2.2 Condensation and evaporation rate

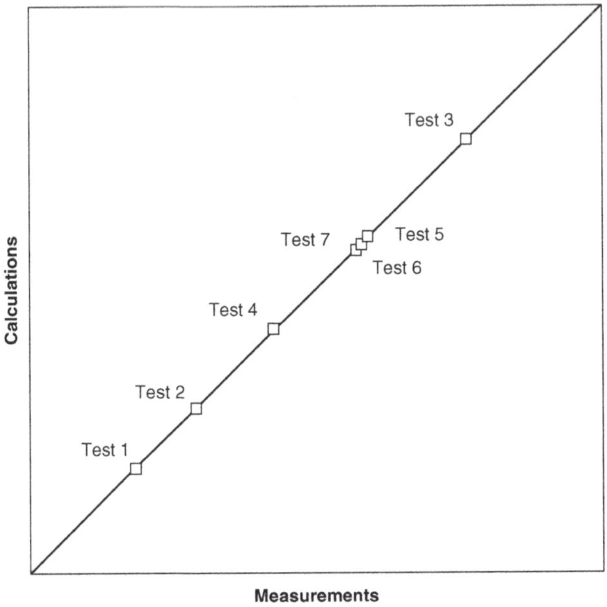

Fig. 7.33. Calculated and measured condensation rate.

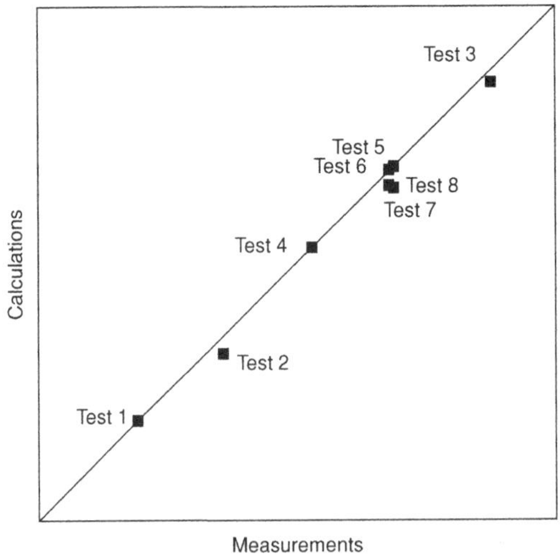

Fig. 7.34. Calculated and measured evaporation rate.

7.2.2.3 Air partial pressure and containment pressure

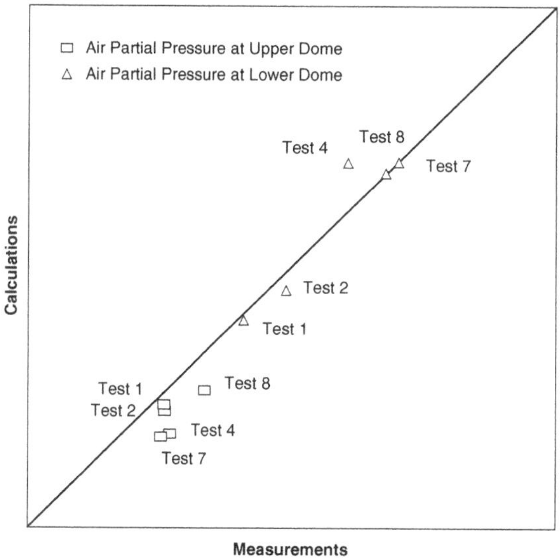

Fig. 7.35. Calculated and measured air partial pressure at upper and lower domes.

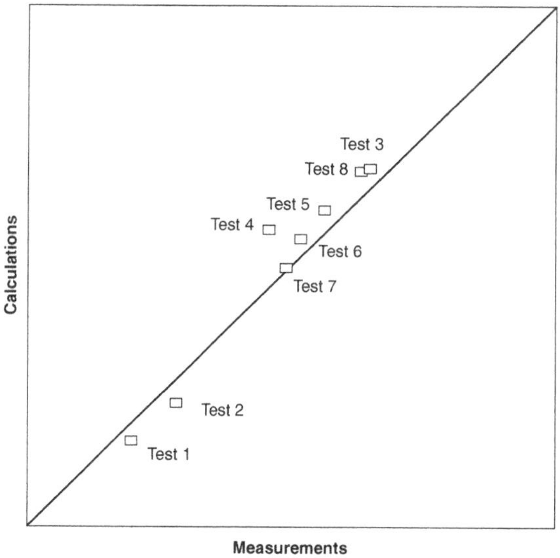

Fig. 7.36. Calculated and measured air containment pressure.

7.2.2.4 Condensation and evaporating film thickness

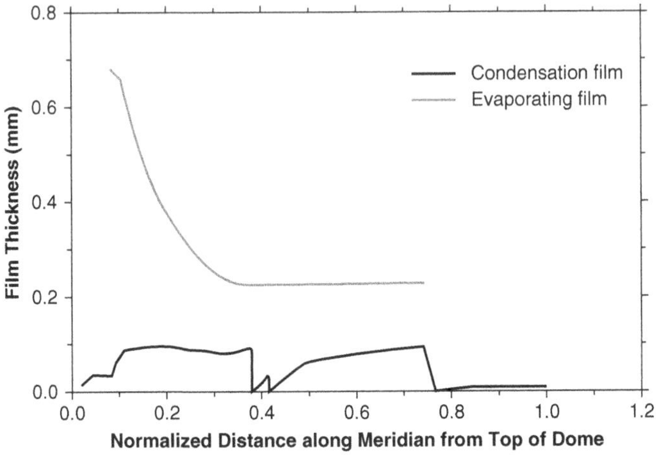

Fig. 7.37. Streamwise variation of condensation and evaporating film thickness at $J = 4$ of test 1.

7.2.2.5 Temperature distributions at various locations

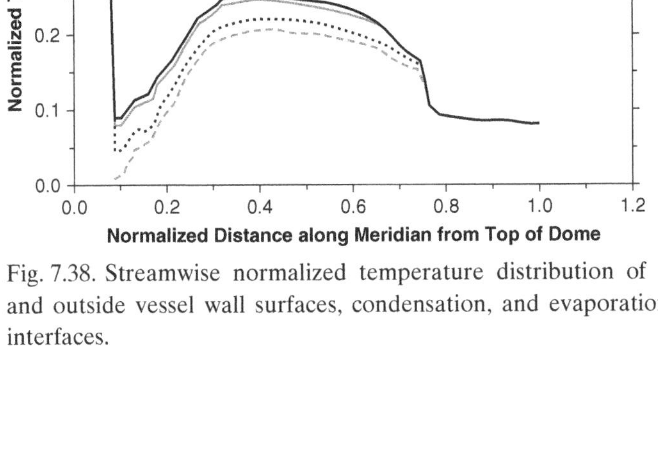

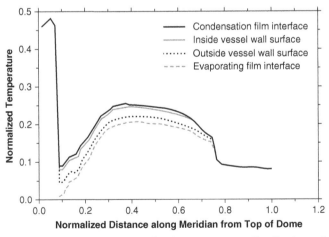

Fig. 7.38. Streamwise normalized temperature distribution of inside and outside vessel wall surfaces, condensation, and evaporation film interfaces.

In summary, we have successfully modeled the complex AP-600 PCCS. We note that the PCCS uses phase change (condensation and evaporation) to remove the heat generated during a DBA from the containment vessel to the environmental via natural convections. At first glance, the modeling of PCCS is a multiphase problem; however, the modeling can be greatly simplified by reducing the multiphase problem into a single phase with multicomponent; see [51]. Both condensation and evaporation rates can be treated as boundary conditions that can be updated iteratively from the temperature of the bulk gaseous mixture for condensation to form liquid film inside the steel containment wall and liquid film surface temperature for evaporation of liquid film for the outside steel containment vessel wall. The advantages of this ingenious formulation [51] are that both the convergence problem and the long computer running time encountered in the multiphase problem can be avoided. The calculated results from the COMMIX code were compared with corresponding experimental data [52], as shown in Figs. 7.29–7.38, with good agreement. Once the COMMIX code is validated with the experimental data, it can carry out sensitivity analysis for further improving the design and performance of the PCCS during normal and off-normal reactor operating conditions.

7.3 Conclusion

From the preceding analyses of Cases (1), (2), (6), and (7), the capability of the COMMIX code based on novel porous media formulation for single-phase applications is clearly demonstrated (1) to compute detailed microflow fields in Von Karmann vortex shedding analysis [Case (1)] and shear-driven cavity flows analysis [Case (2)], and (2) for capturing essential macroflow fields and temperature distribution involving complicated physical phenomena and complex systems, namely, natural convection and flow stratification in a Westinghouse four-Loop PWR [Case (6)] and evaporation of liquid film outside steel containment vessel wall and condensation of steam of steam-air mixture to form a liquid film inside steel containment vessel wall for removing the heat generated during a postulated accident via the PCCS of AP-600 or 1000 reactor [Case (7)]. Comparison of the calculated results from the COMMIX code with corresponding experimental data of Cases (1), (2), (6), and (7) shows good agreement.

To date, nobody has ever solved the multiphase flow conservation equations of mass, momentum, and energy in differential-integral form as derived in this book. Based on the numerical results obtained for single phase and single phase with multicomponent presented in this chapter, we are looking forward with optimism to seeing similar quality results that can be delivered by the novel porous media formulation for two-phase and multiphase flows.

8 Discussion and concluding remarks

Multiphase flows consist of interacting phases that are dispersed randomly in space and in time. It is important to recognize that these turbulent, randomly dispersed multiphase flows can be described only statistically or in terms of averages [1]. Averaging is necessary because of a wide range of sizes, shapes, and densities of dispersed phases and to avoid solving a deterministic multiboundary value problem with the positions of interface being a priori unknown.

In most engineering applications, all that is required is to capture the essential features of the system and to express the flow and temperature field in terms of local-volume or time-volume-averaged quantities while sacrificing some of the details. This book presents the novel porous media formulation for multiphase flow conservation equations, which is an attempt to achieve this goal.

8.1 Time averaging of local volume-averaged phasic conservation equations

The starting point of the formulation is the well-established phasic conservation equations (Navier-Stokes equations) of mass, momentum, and energy with stationary and solid internal structures and their interfacial balance relations. Local volume averaging is performed on the conservation equations and interfacial balance relations, followed by time averaging; this process is named time-volume averaging. In most practical engineering applications of multiphase flow systems, stationary solid internal structures are always associated with the systems. For this reason, the stationary solid structures are included in the derivation via averaging processes for the multiphase conservation equations, so that fluid–structure interactions are an integral part of the system and explicitly taken into account.

Time-volume averaging is chosen over the other forms of averaging for the following three reasons:

1. Local volume averaging has been successfully applied to many laminar, dispersed multiphase flows, and its local volume-averaging theorems are well established and theoretically sound. Because we are concerned here with turbulent multiphase flows in general and dispersed multiphase flows in particular, it is logical to follow by time averaging.
2. Much of our instrumentation, such as the Bourdon tube pressure gauge, hot wire anemometer, and gamma beam, record a space (volume, surface, point) average

followed by a time average. We believe that the dependent variable calculated from time-volume-averaged conservation equations are more simply related to the corresponding variables measured by experiments.

3. Simple time averaging leads to the fractional residence time of a phase rather than volume fraction of a phase. The fractional residence time of a phase becomes equal to the physical volume fraction only in the case of one-dimensional uniform motion of incompressible phases. The thermodynamic properties of a fluid, such as density and specific heat, are cumulative with volume fraction but not with fraction of residence time. Therefore, local volume averaging must precede time averaging (see Chapter 6).

8.1.1 Length-scale restriction for the local volume average

Local volume averaging of conservation equations of mass, momentum, and energy for a multiphase system yields equations in terms of local volume-averaged products of density, velocity, pressure, energy, stresses, and field forces, together with interfacial transfer integrals. These averaging relations are subject to the following length-scale restrictions:

$$d \ll \ell \ll L, \tag{2.4.3}$$

where d is a characteristic length of the pores or dispersed phases, ℓ is a characteristic length of the averaging volume, and L is a characteristic length of the physical system.

Before the local volume-averaged conservation equations can be used either for further analysis or for numerical computation, it is necessary to (1) express the volume averages of the product of the dependent variables in terms of the product of their volume averages, and (2) evaluate the interfacial transfer integrals that depend on the *local* values of the dependent variables at every point on the interface. To this end, it is postulated that a point-dependent variable ψ_k for phase k can be expressed as the sum of its local intrinsic volume average ${}^{3i}\langle\psi_k\rangle$ and a spatial deviation $\tilde{\psi}_k$. ψ_k can be a scalar, vector, or tensor. Both ${}^{3i}\langle\psi_k\rangle$ and $\tilde{\psi}_k$ have a low-frequency component to be denoted by the subscript LF and a high-frequency component to be denoted by a prime. Thus, solutions of local volume-averaged conservation equations call for expressing these local volume-averaged products in terms of products of averages, and interfacial transfer integrals can be evaluated by introducing spatial deviation.

In nonturbulent flows, this can be achieved by expressing the "point" variable as the sum of its intrinsic volume average and a spatial deviation.

8.1.2 Time scale restriction in the time averaging

In turbulent flows, the same can be achieved through subsequent time averaging over a duration T such that

$$\tau_{HF} \ll T \ll \tau_{LF}, \tag{5.1.4}$$

where τ_{HF} is a characteristic time of high-frequency fluctuation as defined as

$$\tau_{HF} = \frac{L_C}{(\Delta U)_C} = \frac{\text{characteristic dimension of physical systems}}{\text{characteristic low-frequency speed variation at a typical location}} \tag{5.1.3a}$$

and τ_{LF} is a characteristic time of low-frequency fluctuation as defined as

$$\tau_{HF} = \frac{\Lambda}{(rms\ U')} = \frac{\text{characteristic length scale of high-frequency fluctuation}}{\text{root mean square of fluctuation velocity or turbulence intensity}} = \frac{1}{\text{characteristic spectral frequency}}. \tag{5.1.3b}$$

In this case, an instantaneous "point" variable ψ_k of phase k is decomposed into a low-frequency component ψ_{kLF} and a high-frequency component ψ'_k, similar to Reynolds analysis of turbulent flow. The low-frequency component consists of the sum of the local intrinsic volume average ${}^{3i}\langle\psi_k\rangle_{LF}$ and its local spatial deviation $\tilde{\psi}_{kLF}$. Time averaging then reduces the volume-averaged products to products of averages plus terms representing, in general, eddy and dispersive diffusivities of mass, Reynolds and dispersive stresses, and eddy and dispersive conductivities of heat. These terms arise from both high-frequency fluctuations and local spatial deviations. This procedure of time averaging after local volume averaging leads to a set of differential-integral equations of conservation for multiphase flow. This set of multiphase flow conservation equations is particularly suitable for numerical analysis with a staggered grid computational system (Appendix A).

8.1.3 Time-volume-averaged conservation equations are in differential-integral form

When we presented the papers from 1980 to 1984 [1–5], we knew then that the conservation equations of mass, momentum, and energy for multiphase flows should be in differential-integral form, but we did not know how to extend the single-phase to multiphase formulation with turbulence until we published another paper in 2007 [6] and now presented in detail in this book.

A set of conservation equations of mass, momentum, total energy, internal energy and enthaply of a multiphase flow systems (phase $k, f\ldots$) with internal stationary structure (phase w) via time-volume-averaging of Navior-Stoke equations represented by Eqs. (5.3.4), (5.3.3c), (5.5.7f), (5.5.7g), (5.5.7j), (5.7.10), (5.9.7), and (5.11.10), respectively, has been rigorously derived for the first time. The corresponding time-volume-averaged interfacial balance equations (kf, wk, fk, $wf\ldots$) are taken from the derived conservation of mass, momentum, total energy, internal energy and enthalpy equation, represented by Eqs. (5.4.1), (5.6.1), (5.8.1), (5.10.1), and (5.12.1), respectively. These equations, named Sha's Multi-phase Flow Equations, are applicable to both dispersed and separated flows [58], with zero thickness of phase interfaces. The conservation equations are in differential-integral form, in contrast to a set of partial differential equations used currently. The integrals arise due to interfacial mass, momentum, and energy transfer. Although scientists and engineers have been working on two-phase problems since the invention of the steam engine and boiler almost a century ago, a correct set of conservation equations for design and performance analysis of two-phase and multiphase flow systems has never been established. The derived equations presented here serve as a reference point for modeling multiphase flow with simplified approximations. Moreover, they provide theoretical guidance and physical insights useful for developing correlations for quantifying interfacial mass, momentum, and energy transfer between phases.

8.1.4 Unique features of time-volume-averaged conservation equations

Careful examination of the derived time-volume-averaged conservation equations of multiphase flows in this book indicates the following unique features:

1. The time-volume-averaged interfacial mass generation rate of phase k per unit volume in υ, as shown in Eq. (5.3.3c), is carried over directly into both time-volume-averaged momentum [Eq. (5.5.7f)] and energy equations {total energy [Eq. (5.7.10)], internal energy [Eq. (5.9.7)], and enthalpy [Eq. (5.11.10)]}, which are intuitively expected.

$$\gamma_\upsilon \alpha_k {}^t\langle \Gamma_k \rangle = -\upsilon^{-1} \int_{A_k} {}^t\langle \rho_k(\underline{U}_k - \underline{W}_k)\rangle \cdot \underline{n}_k dA$$
$$= \gamma_\upsilon {}^{3i}\langle \rho_k \rangle_{LF} \left(\frac{\partial \alpha_k}{\partial t} + {}^{3i}\langle \underline{U}_k \rangle_{LF} \cdot \nabla \alpha_k \right)$$
$$+ (MTI)_k \qquad (5.3.3c)$$

2. The time-volume-averaged continuity equations with constant density are shown in Eqs. (5.3.5b) and (5.3.5c).

$$\gamma_A \nabla \cdot \alpha_k {}^{2i}\langle \underline{U}_k \rangle_{LF} = \gamma_v {}^{3i}\langle \underline{U}_k \rangle_{LF} \cdot \nabla \alpha_k + {}^{o}\langle MTI \rangle_k \quad (5.3.5b)$$

$$^{o}\langle MTI \rangle_k = -v^{-1} \int_{A_k} \underline{U}_{kLF} \cdot \underline{n}_k dA \quad (5.3.5c)$$

For a single-phase system without internal structures, $\alpha_k = 1$, $A_k = 0$, $\gamma_v = \gamma_A = 1$, Eqs. (5.3.5b) and (5.3.5c) will reduce to

$$\nabla \cdot \underline{U}_k = 0. \quad (5.3.5d)$$

This is the expected result.

3. The time-volume-averaged linear momentum equations are shown in Eqs. (5.5.7f), (5.5.7g), and (5.5.7j). For a single-phase system without internal structures, where the system is at rest, $\alpha_k = 1$, $A_k = 0$, $\gamma_v = \gamma_A = 1$, and quantities associated with $\underline{U}_k$ vanish. Equations (5.5.7f), (5.5.7g), and (5.5.7j) will reduce to

$$-\nabla P_k = \rho_k \underline{g}, \quad (5.5.10a)$$

where $\underline{g}$ is the gravitational acceleration and Eq. (5.5.10a) satisfies the basic relation of fluid hydrostatics.

4. The time-volume-averaged interfacial mass, momentum, and energy (total energy, internal energy and enthalpy balance) equations provide detailed mathematical expressions as shown in Chapter 5, Sec. 5.14, for the first time. By evaluating these mathematical expressions, they give physical insights and relative importance of each term or mechanism of the system under the consideration without any guessing.

5. We note that the similarities between the final set of time-volume-averaged internal energy and enthalpy equations are amazing. Moreover, the three energy equations – namely, total energy, internal energy, and enthalpy – are not independent of one another. Based on the formulation presented here, as expected, the formulation of the total energy is relatively more complicated; the complexity of the formulations based on internal energy and enthalpy are comparable, but the formulation of internal energy is slightly simpler.

The above unique features give confidence in the derived time-volume-averages multiphase conservation equations.

8.2 Novel porous media formulation

In most engineering problems, the external geometry of physical system under consideration and the internal complex structure involved do not, in general, lend themselves to one of Cartesian, cylindrical, or spherical coordinates. The novel porous media formulation has the flexibility to model the external irregular physical boundaries, as well as the internal complex structure shape and size, into a selected coordinates by judiciously using volume porosity, directional surface porosities, distributed resistance, and distributed heat source and sink [see Cases (6) and (7) in Chapter 7].

The *volume porosity* is defined as the ratio of volume occupied by the fluids in a control volume to the control volume. *Directional surface porosities* are defined as a free-flow area of a control surface to the control surface area in three principal directions that are readily calculable quantities [2–6], which is derived naturally through the local volume-averaged conservation equations. Most practical engineering problems involve many complex shapes and sizes of solid internal structures whose distributed resistance is impossible to quantify accurately. This eliminates the sole reliance on empirical estimate of distributed resistance that is impossible to accurately quantify for the many complex structures often involved in the analyses. It also facilitates mockup of numerical simulation models to real engineering systems (see Chapter 7). The directional surface porosities greatly improve the resolution and modeling accuracy. The incorporation of spatial deviation into the instantaneous point-dependent variables made it possible to evaluate interfacial transfer integrals.

In numerical simulations of practical engineering problems, the computational mesh setup may not perfectly match the size and shape of the stationary solid structures, such as perforated baffles and nozzle of a large tank, located in a computational domain. Without directional surface porosities, the calculated velocities passing through these structures will be either too high or too low, resulting in erroneous mass, momentum, and energy transfer.

We note that, for the conservation equations presented in this book, the local averaging volume is unrelated to the volume of a computational cell used in the numerical computations.

8.2.1 Single-phase implementation

In many practical engineering analyses, the computational domain can be subdivided into many subdomains. Among these subdomains, some are more suitable to use the conventional porous media formulation (e.g., nuclear reactor core) and others are appropriate to use the continuum formulation (e.g., nuclear reactor upper plenum, where there is no solid structures) [3]. By appropriate input of the volume porosity, directional surface porosities, distributed resistance, and distributed heat source and sink for each computation cell, the novel porous media formulation enabled the solution of real world challenging engineering problems with many stationary and complex internal structures.

The novel porous media formulation for single-phase flows was implemented in the generic three-dimensional, time-dependent family of COMMIX codes [37]. The COMMIX suite with this formulation has been widely adopted by the nuclear reactor community in the United States and abroad, including Germany, France, the UK, Italy, Finland, Japan, China, and South Korea [6], and distributed and used in U.S. universities [53], utility companies [43,44], and industries [45,46].

For example, the code has been used for pressurized thermal shock (PTS) analysis, which helped EPRI resolve the PTS issue [43,44]; heat exchanger analysis is to help heat exchanger industry improving their products and design [45,46]; in-vessel analysis of coupling effects of reactor components to improve and enhance reactor performance and safety [47,48]; analysis of natural convection phenomena in a prototypic PWR during a postulated degraded core accident will enhance the understanding of natural convection patterns and their impact on accident progression and consequences [49,50]; and analysis of a PCCS leads further improvement in design and performance of PCCS [51,52]. All the results of these analyses were compared with corresponding experimental data, and the agreement was good.

Equally important are the benefits derived from the novel porous media formulation in areas beyond applications to nuclear reactor systems. Significant spin-offs have been observed in the simulation of casting processes [54], fluidized bed combustor modeling [55], solar energy applications [56], and flotation applications for coal precleaning [57].

8.2.2 Multiphase flow

To date, computational experiences with the novel porous media formulation have focused mainly on single phase and single phase with multicomponent applications. Based on the success of these applications presented in Chapter 7 and Section 8.2.1, and described elsewhere [2–6], the novel porous media formulation already represents a flexible and unified formulation for studying and applications of computational fluid dynamics with heat transfer.

This book makes two major contributions. First, the conservation equations previously used were approximated as a set of partial differential equations. Here, however, the multiphase flow equations have been rigorously derived and presented in *differential-integral* form. Second, the recent rigorous theoretical extension of the novel porous media formulation to *multiphase* flows, as described in this book, has far-reaching implications and benefits for research and development of multiphase flows for both academic and industry.

We emphasize that these equations presented in this book have never been employed for two and multiphase flow analysis. The next step is for researchers to implement this new set of differential-integral equations in a parallel processors program, taking advantage of large-scale computers to address challenging real world problems in science and engineering.

8.3 Future research

Because of the length-scale restrictions, the time-volume-averaged conservation equations for multiphase flow are strictly valid for highly dispersed systems. When these equations are applied to systems (1) that are not highly dispersed; (2) whose relative magnitude of α', v', and A' are not negligible; and (3) whose time scale in inequalities as shown in Eq. (5.1.4) are not satisfied, or whose low-frequency and high-frequency quantities are not separable, the extent and nature of errors involved remain a subject of further research.

The analysis of multiphase flow calls for the solution of the time-volume-averaged differential-integral equations of both conservation and their interfacial balance equations, which are developed in Chapter 5. These equations, coupled with appropriate initial and boundary conditions, are applicable to both dispersed and separated flows [58]. For the nonturbulent flows, the time-volume-averaged conservation equations and their interfacial balance equations can readily be obtained by dropping the high-frequency terms.

An examination of these equations reveals immediately that they are incomplete in that constitutive relations for the diffusive, dispersive, turbulent, and interfacial transfer need to be developed. Collectively, this constitutes the closure problem. We note that the integrand of the interfacial integrals consists of the local values of the dependent variables. Equivalently, it contains the deviation of the local value of the variable from its intrinsic local volume average and, in the presence of high-frequency fluctuations, its turbulent component. The closure problem is not unlike that in the analysis of turbulent flow, but with additional complications. In the absence of turbulence, a closure scheme for the determination of the spatial deviation of the dependent variable for systems involving diffusion and first-order chemical reaction is given by Crapiste et al. [59]. A rigorous approach to treat the general closure problem, including convective transport and turbulence, will, no doubt, remain a challenge [6].

At the present time, the evaluation of the interface transfer integrals in the time-volume-averaged conservation equations is not generally known. An order-of-magnitude analysis to assess the relative importance of these interface transfer integrals would be helpful. One of the fundamental problems in understanding multiphase flow is the lack of knowledge of mass, momentum, and energy transfer at the interface. In the past, empirical correlations were developed from experimental data to quantify interfacial mass, momentum, and energy transfer rates, often without sound theoretical basis. Therefore, these correlations are valid only in the range of operating conditions for which the experimental data are obtained. Other urgently needed information is the quantification of transport properties such as eddy and dispersive diffusivities of mass, Reynolds and dispersive stresses, and eddy and dispersive conductivities of heat by performing planned experiments in conjunction with analysis.

The current situation in the area of multiphase flows with heat transfer is confusing. This confusion arises from two factors: (1) the failure to recognize that the turbulent, dispersed multiphase flows can be described only in terms of averages or statistically in the early development of multiphase flow [1]; and (2) the fact that because constitutive relations vary with different sets of conservation equations, we have been using many different sets of conservation equations resulting from different formulation equations for multiphase flow. Unless the same set of conservation equations is used in the multiphase flow research community, this confusion will persist. We propose that the novel porous media formulation leads the way in becoming a universally accepted formulation for multiphase conservation equations because it has the following unique features.

Theoretically sound. The formulation starts with the local volume averaging of the Navier-Stokes equations and interfacial balance relations of each phase via local volume averaging theorems that are well established and theoretically sound.

Time averaging is then performed on both the local volume-averaged conservation equations and interfacial balance relations. The time-volume-averaged conservation equations are in differential-integral form and are in contrast to the currently used set of partial differential equations.

Generic. The formulation has the flexibility to include or not to include the high-frequency fluctuation variables of α_k' (high-frequency fluctuation of the volume fraction), υ_k' (high-frequency fluctuation of the local averaging volume), and A_k' (high-frequency fluctuation of the interfacial area), based on the need from physics of the problem under consideration.

Consistent. The multiphase conservation equations are derived based on a single formulation. For example, if we need to retain α_k', A_k', or υ_k', then we can follow the same procedure to derive another set of multiphase flow conservation equations based on the same formulation. Because α_k', A_k', or υ_k' have very close intertwining relations, careful consideration must be given so that consistence can be maintained, and the current status of understanding of phase interface must be validated with pertinent experimental data (see Appendix E).

The benefits resulted from the idea of a single, universally accepted formulation to derive multiphase flow conservation equations are enormous. The effort and expense in developing constitutive relations will no longer need to be repeated for each specific problem. Furthermore, the range of interest in operating conditions or dependent variables for engineering problems varies among research groups; a single formulation will enable the constitutive relations developed by these different groups to be integrated, leading to a better understanding of the class of the problems.

At the present time, the interface thickness between two phases is assumed to be zero for simplicity. With advances in instrumentation in measuring interface area, it is possible for removal of the assumption of zero interfacial thickness. This is undoubtedly a challenging area.

In a previous study of turbulence in single-phase fluid, a third component, the coherence component, was added [60]. At the present stage of development of the mechanics of multiphase flow, we should seriously consider this option because it can be verified or observed.

Recent publications on multiphase flows [58, 61, 62] are very good reference books, but they do not deal with the conservation equations of multiphase flows. It was cited in Ref. [61] that "the heterogeneous porous media formulation presented by Sha and Chao [6] is to date the effective concept for deriving multiphase flow equations applicable in complicated technical facilities." We believe strongly that not having correct conservation equations for multiphase flow is equivalent to building a house without good foundation. This book is intended to bring engineers' and scientists' attention to the subject. Progress in science is a step function, and this is the first step. It is hoped that this book will chart the direction of future research and development of multiphase flow.

8.4 Summary

A set of rigorously derived multiphase flow conservation equations in a region containing stationary and dispersed solid structures via time-volume-averaging has been presented. This set of conservation equations is in differential-integral form, in contrast to a set of partial differential equations used currently. We hope and encourage researchers to use the set of conservation equations described in this book as a reference point for modeling multiphase flow, and these equations provide theoretical guidance and physical insights for both planning experimental facility and developing a three-dimensional, time-dependent computer code with parallel processors. Iteration among the theory, experimental data, and computer modeling are necessary to speed up the development and validation of closure relations for quantifying interfacial mass, momentum, and energy transfer between phases.

Introduction of the concept of directional surface porosities [2–6] is one of the keys to the success of the novel porous media formulation and greatly improved the resolution and modeling accuracy. The incorporation of spatial deviation into the instantaneous point-dependent variables made it possible to evaluate all interfacial transfer integrals. The novel porous media formulation represents a new, flexible, and unified approach to computational fluid dynamics with heat transfer for multiphase flows and made it possible to solve real challenge engineering problems with stationary and complex solid structures.

The intended goal of this book is to make the novel porous media formulation for multiphase flow conservation equations equivalent to the Navier-Stokes equations for single phase. The idea is to establish a universally accepted formulation for multiphase flow conservation equations that will greatly reduce both the efforts and the expenses involved in developing the constitutive relations. Most important, we hope that the novel porous media formulation will enhance and facilitate scientific understanding of multiphase flows.

APPENDIX A

Staggered-grid computational system

A staggered-grid computational system is widely used in numerical analysis. A family of COMMIX codes [37] is no exception; the codes greatly benefited by using this system. In this system, all dependent nonflow variables (pressure, temperature, density, total energy, enthalpy, internal energy, mass fraction of phases, turbulent transport quantities, thermal physical properties, etc.) are calculated for the computational cell center (I, J, K) in three-dimensional orthogonal coordinates, and all flow variables (velocity components) are calculated for the surfaces of the computational cell ($I \pm 1/2$, $J \pm 1/2$, $K \pm 1/2$). The computational cell is defined by the locations of cell volume faces, and a grid point is placed in the geometric center of each cell volume; cell sizes can be nonuniform. This type of construction of computational system is shown in Fig. A.1.

Consider the control volume for a nonflow variable shown in Fig. A.2. It is constructed around a grid point 0, which has grid points 1 $(i - 1)$ and 2 $(i + 1)$ as its west and east neighbors, 3 $(j - 1)$ and 4 $(j + 1)$ as its front and rear

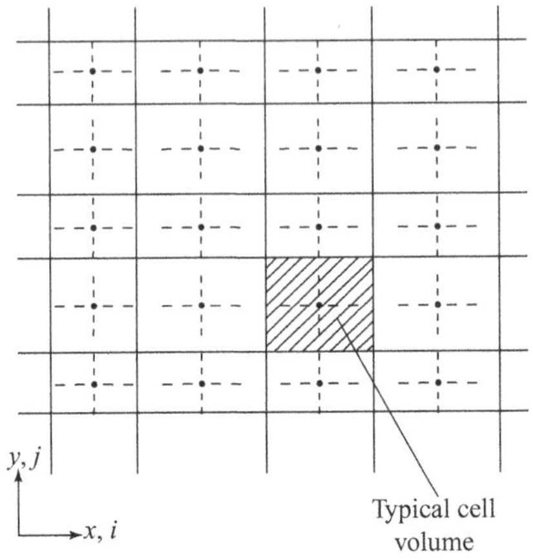

Fig. A.1. Construction of cell volumes.

neighbors, and 5 $(k-1)$ and 6 $(k+1)$ as its south and north neighbors.

Although most dependent variables are calculated for a grid point, the velocity components u, v, and w are exceptions. They are calculated not for a grid point but for displaced or staggered locations of the velocity components such

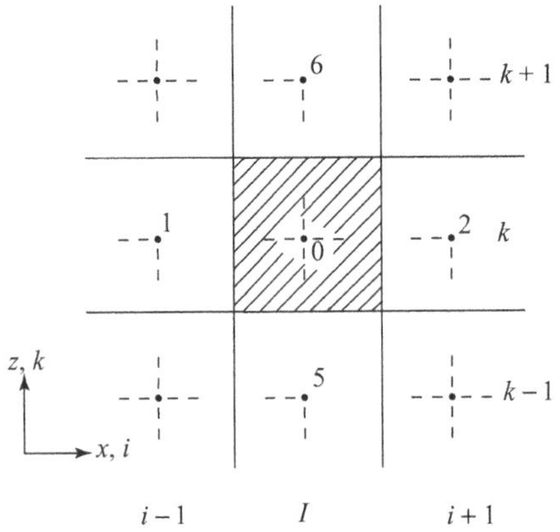

Fig. A.2. Cell volume around point 0 in $i, j, k,$ notation.

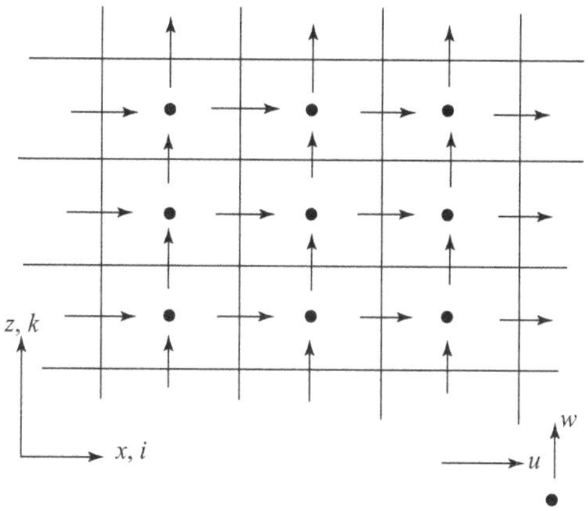

Fig. A.3. Staggered grid.

that they are placed on the faces of a control volume or cell volume, as shown in Fig. A.1. Thus, the k-component velocity w is calculated at the faces normal to the k direction.

Figure A.3 shows the locations of u and w by short arrows on a two-dimensional grid; the three-dimensional counterpart can be easily visualized. Relative to a grid point, the u location is displaced only in the i direction, the w location only in the k direction, and so forth. The location for w thus lies in the k-direction link, joining two adjacent grid points. The pressure difference between these grid points will be used to drive the velocity w located between them. This is the main feature of the staggered grid.

A direct consequence of the staggered grid is that the control volume to be used for the conservation of momentum must also be staggered. The control volume shown in Figs. A.1 and A.2 is referred to as the main control volume. The control volume for momentum is staggered in the

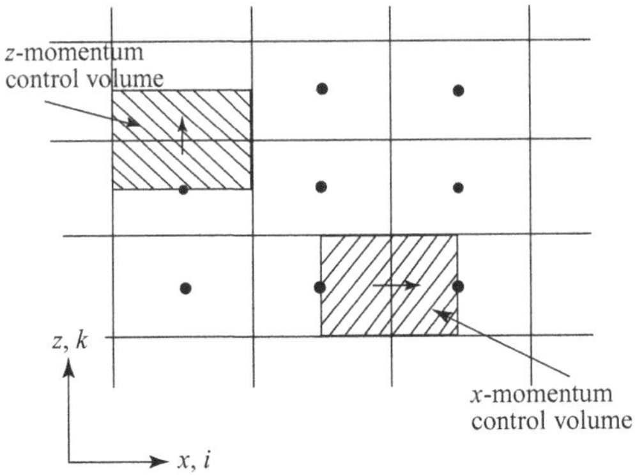

Fig. A.4. Momentum control volumes.

direction of the momentum so that the faces normal to that direction pass the grid point (Fig. A.4). Thus, the pressures at these grid points can be directly used to calculate the pressure force on the momentum control volume. Figure A.4 shows the momentum control volume for the x and z directions.

The formulation of multiphase flow equations as presented here is closely related to the staggered-grid computational system. For example, the physical meaning of the second and third term on the LHS of time-volume-averaged continuity [Eq. (5.3.4)] and the first and second terms on the LHS of time-volume-averaged momentum [Eq. (5.5.7f)] illustrate the salient features of the formulation:

$$\gamma_A \nabla \cdot {}^{t}\langle \alpha_k {}^{2i}\langle \rho_k \underline{U}_k \rangle\rangle = \gamma_A \nabla \cdot \left(\alpha_k {}^{2i}\langle \rho_k \rangle_{LF} {}^{2i}\langle \underline{U}_k \rangle_{LF} + \underline{\psi}^{2i}_{mk}\right), \tag{A.1}$$

where $\underline{\psi}^{2i}_{mk} = \alpha_k {}^{2i}\langle \tilde{\rho}_{kLF} \underline{\tilde{U}}_{kLF} \rangle + \alpha_k {}^{t2i}\langle \rho'_k \underline{U}'_k \rangle$.

All variables are evaluated at the surface of the computation cell under consideration.

Because ${}^{3i}\langle \rho_k \rangle$ is calculated at the cell center, which is not available, an appropriate averaging of the neighboring cell densities of the surface are used. The alternative is to use upwind density knowing the ${}^{2i}\langle \underline{U}_k \rangle$ of the surface:

$$\gamma_v \frac{\partial}{\partial t} {}^t\langle \alpha_k {}^{3i}\langle \rho_k \underline{U}_k \rangle \rangle = \gamma_v \frac{\partial}{\partial t} (\alpha_k {}^{3i}\langle \rho_k \rangle_{LF} {}^{3i}\langle \underline{U}_k \rangle_{LF} + \underline{\psi}^{3i}_{mk}), \tag{A.2}$$

where $\underline{\psi}^{3i}_{mk} = \alpha_k {}^{3i}\langle \tilde{\rho}_k \underline{\tilde{U}}_k \rangle + \alpha_k {}^{t3i}\langle \rho'_k \underline{U}'_k \rangle$.

All variables are evaluated at the center of computational cell under consideration.

Because ${}^{3i}\langle \underline{U}_k \rangle$ is calculated at the surface of the cell, which is not available, an appropriate averaging of the neighboring surface velocities of the cell are used. The alternative is to use so-called donor flow formulation [63].

The advantage of this formulation is that the precise locations of variables to be evaluated are clearly indicated.

Generally speaking, staggered-grid computational system is adequate for most engineering applications, however, in some situations such as external and internal boundary conditions require special consideration the boundary-fitted coordinates transformation [64,65,66] can be considered.

APPENDIX B

Physical interpretation of $\nabla \alpha_k = -\upsilon^{-1} \int_{A_k} \underline{n}_k dA$ with $\gamma_\upsilon = 1$ (B.1)

To provide a physical interpretation of Eq. (B.1) [36], which is Eq. (2.4.9) with $\gamma_\upsilon = 1$, we consider a dispersed system and an averaging volume in the shape of a rectangular parallelepiped $\Delta x \Delta y \Delta z$ with its centroid located at (x, y, z), as illustrated in Fig. B.1a. Its top view is shown Fig. B.1b.

Clearly, for those elements of the dispersed phase k that are completely inside the averaging volume,

$$\int_{\delta A_k} \underline{n}_k dA_k = 0, \tag{B.2}$$

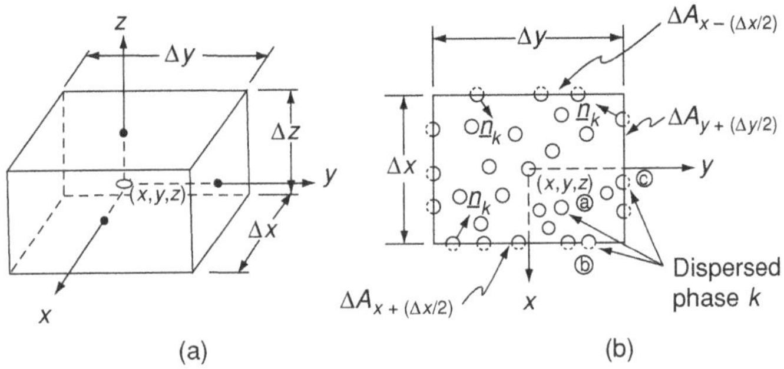

Fig. B.1. Physical interpretation of Eq. (2.4.9) for $\gamma_\upsilon = 1$.

where δA_k is the closed surface of the element. Such an element, labeled ⓐ in Fig. B.1b, may be a bubble or a droplet, spherical or nonspherical. Next, we consider those elements of the dispersed phase that are intersected by the boundary surface $\Delta A_{x+(\Delta x/2)}$. One such element is labeled ⓑ in Fig. B.1b. The unit outdrawn normal vector $\underline{n}_k$ can be represented by

$$\underline{n}_k = \underline{i}e_1 + \underline{i}e_2 + \underline{k}e_3, \tag{B.3}$$

where i, j, and k are unit vectors pointing in the positive directions of x, y, and z-axis, respectively, and e_1, e_2, and e_3 are the direction cosines of $\underline{n}_k$. If we denote the portion of the interfacial area of element ⓑ that is inside the averaging volume v by $\delta A_{k,[x+(\Delta x/2)]}$, and its area of intersection with the surface $\Delta A_{x+(\Delta x/2)}$ by $\delta A_{k,x+(\Delta x/2)}$, then

$$\int_{\delta A_{k,[x+(\Delta x/2)]}} e_1\, dA = \delta A_{k,x+(\Delta x/2)} \tag{B.3a}$$

and

$$\int_{\delta A_{k,[x+(\Delta x/2)]}} e_2\, dA = \int_{\delta A_{k,[x+(\Delta x/2)]}} e_3\, dA = 0. \tag{B.3b,c}$$

Likewise, for an element of phase k that is intersected by the surface $\Delta A_{x-(\Delta x/2)}$, we have

$$\int_{\delta A_{k,[x-(\Delta x/2)]}} e_1\, dA = \delta A_{k,x-(\Delta x/2)} \tag{B.4a}$$

and

$$\int_{\delta A_{k,[x-(\Delta x/2)]}} e_2\, dA = \int_{\delta A_{k,[x-(\Delta x/2)]}} e_3\, dA = 0. \quad \text{(B.4b,c)}$$

Following the same procedure, we have for an element of phase k that is intersected by the surface $\Delta A_{y+(\Delta y/2)}$ [labeled ⓒ in Fig. B.1b]:

$$\int_{\delta A_{k,[y+(\Delta y/2)]}} e_2\, dA = \delta A_{k,\, y+(\Delta y/2)} \quad \text{(B.5a)}$$

and

$$\int_{\delta A_{k,[y+(\Delta y/2)]}} e_1\, dA = \int_{\delta A_{k,[y+(\Delta y/2)]}} e_3\, dA = 0, \quad \text{(B.5b,c)}$$

where $\delta A_{k,[y+(\Delta y/2)]}$ denotes the portion of the interfacial area of the element ⓒ that is inside υ, and its intersection by the surface $\Delta A_{y+(\Delta y/2)}$ is $\delta A_{k,y+(\Delta y/2)}$. Similar expressions can be written for elements of phase k that are intersected by the bounding surface $\Delta A_{y-(\Delta y/2)}$, $\Delta A_{z+(\Delta z/2)}$, and $\Delta A_{z-(\Delta z/2)}$.

The x component of the integral on the RHS of Eq. (B.1) is

$$\left(-\upsilon^{-1}\int_{A_k} \underline{n}_k\, dA\right)_x = -\frac{1}{\Delta x \Delta y \Delta z}\left(-\sum \delta A_{k,x+(\Delta x/2)} + \sum \delta A_{k,x-(\Delta x/2)}\right), \quad \text{(B.6)}$$

where the summation is taken for all elements of phase k cut through by the bounding surfaces $\Delta A_{x+(\Delta x/2)}$ and $\Delta A_{x-(\Delta x/2)}$. Using the relationship given by Eq. (2.3.8a), one has

$$\frac{\sum \delta A_{k,x+(\Delta x/2)}}{\Delta y \Delta z} = \alpha_{k,x+(\Delta x/2)}$$

and

$$\frac{\sum \delta A_{k,x-(\Delta x/2)}}{\Delta y \Delta z} = \alpha_{k,x-(\Delta x/2)}.$$

Thus,

$$\left(-\upsilon^{-1} \int_{A_k} \underline{n}_k \, dA \right)_x = \frac{\Delta \alpha_{k,x}}{\Delta x}, \tag{B.7}$$

where $\Delta \alpha_{k,x} = \alpha_{k,x+(\Delta x/2)} - \alpha_{k,x-(\Delta x/2)}$. Similar expressions can be written for the y and z component of the indicated integral. As has been pointed out in Section 2.3 for highly dispersed systems, $\Delta \alpha_{k,x} = \Delta \alpha_{k,y} = \Delta \alpha_{k,z} = \Delta \alpha_k$. It follows then that

$$-\upsilon^{-1} \int_{A_k} \underline{n}_k \, dA = i \frac{\Delta \alpha_k}{\Delta x} + j \frac{\Delta \alpha_k}{\Delta y} + k \frac{\Delta \alpha_k}{\Delta z}, \tag{B.8a}$$

for which we can write

$$\nabla \alpha_k = -\upsilon^{-1} \int_{A_k} \underline{n}_k \, dA \tag{B.8b}$$

in view of the length-scale restrictions of Eq. (2.4.3).

Evaluation of ${}^{t}\langle {}^{3i}\langle \underline{\underline{\tau}}_k \rangle\rangle$ for non-Newtonian fluids with $\gamma_v = 1$

For a Newtonian fluid, the stress and strain rate of a fluid phase k are linearly related and are expressible as

$$\underline{\underline{\tau}}_k = \left(\lambda_k - \frac{2}{3}\mu_k\right)\nabla \cdot \underline{U}_k \underline{\underline{I}} + \mu_k[\nabla, \underline{U}_k + (\nabla, \underline{U}_k)_c], \quad (5.5.4c)$$

in which all quantities have been previously defined. When the viscosity coefficients λ_k and μ_k are dependent on the strain rate [36], they are decomposed in accordance with Eq. (5.1.8):

$$\lambda_k = {}^{3i}\langle \lambda_k \rangle_{LF} + \tilde{\lambda}_{kLF} + \lambda'_k \quad (C.1)$$

$$\mu_k = {}^{3i}\langle \mu_k \rangle_{LF} + \mu'_{kLF} + \mu'_k. \quad (C.2)$$

It is straightforward to demonstrate that

$$\begin{aligned}{}^{3i}\langle \underline{\underline{\tau}}_k \rangle = {}& {}^{3i}\left\langle \lambda_k - \frac{2}{3}\mu_k \right\rangle_{LF} (\nabla \cdot {}^{3i}\langle \underline{U}_k \rangle_{LF})\underline{\underline{I}} \\ & + {}^{3i}\left\langle \lambda_k - \frac{2}{3}\mu_k \right\rangle_{LF} {}^{3i}\langle \nabla \cdot \underline{\tilde{U}}_{kLF} \rangle \underline{\underline{I}} \\ & + {}^{3i}\left\langle \lambda_k - \frac{2}{3}\mu_k \right\rangle_{LF} {}^{3i}\langle \nabla \cdot \underline{U}'_k \rangle \underline{\underline{I}}\end{aligned}$$

$$
\begin{aligned}
&+ {}^{3i}\left\langle\left(\tilde{\lambda}_{kLF} - \frac{2}{3}\tilde{\mu}_k\right)\nabla\cdot\underline{\tilde{U}}_{kLF}\right\rangle\underline{\underline{I}} \\
&+ {}^{3i}\left\langle\left(\tilde{\lambda}_{kLF} - \frac{2}{3}\tilde{\mu}_k\right)\nabla\cdot\underline{U}'_k\right\rangle\underline{\underline{I}} \\
&+ {}^{3i}\left\langle\left(\lambda'_k - \frac{2}{3}\mu'_k\right)\nabla\cdot {}^{3i}\langle\underline{U}_k\rangle_{LF}\right\rangle\underline{\underline{I}} \\
&+ {}^{3i}\left\langle\left(\lambda'_k - \frac{2}{3}\mu'\right)\nabla\cdot\underline{\tilde{U}}_{kLF}\right\rangle\underline{\underline{I}} \\
&+ {}^{3i}\left\langle\left(\lambda'_k - \frac{2}{3}\mu'_k\right)\nabla\cdot\underline{U}'_k\right\rangle\underline{\underline{I}} \\
&+ {}^{3i}\langle\mu_k\rangle_{LF}[\nabla, {}^{3i}\langle\underline{U}_k\rangle_{LF} + (\nabla, {}^{3i}\langle\underline{U}_k\rangle_{LF})_c] \\
&+ {}^{3i}\langle\mu_k\rangle_{LF}\,{}^{3i}\langle\nabla, \underline{\tilde{U}}_{kLF} + (\nabla, \underline{\tilde{U}}_{kLF})_c\rangle \\
&+ {}^{3i}\langle\mu_k\rangle_{LF}\,{}^{3i}\langle\nabla, \underline{U}'_k + (\nabla, \underline{U}'_k)_c\rangle \\
&+ {}^{3i}\langle\tilde{\mu}_{kLF}[\nabla, \underline{\tilde{U}}_{kLF} + (\nabla, \underline{\tilde{U}}_{kLF})_c]\rangle \\
&+ {}^{3i}\langle\tilde{\mu}_{kLF}[\nabla, \underline{U}'_k + (\nabla, \underline{U}'_k)_c]\rangle \\
&+ {}^{3i}\langle\mu'_k\rangle[\nabla, {}^{3i}\langle\underline{U}_k\rangle_{LF} + (\nabla, {}^{3i}\langle\underline{U}_k\rangle_{LF})_c] \\
&+ {}^{3i}\langle\mu'_k[\nabla, \underline{\tilde{U}}_{kLF} + (\nabla, \underline{\tilde{U}}_{kLF})_c]\rangle \\
&+ {}^{3i}\langle\mu'_k[\nabla, \underline{U}'_k + (\nabla, \underline{U}'_k)_c]\rangle. \qquad \text{(C.3)}
\end{aligned}
$$

Subsequent time averaging gives

$$
\begin{aligned}
{}^{t}\langle{}^{3i}\langle\underline{\underline{\tau}}_k\rangle\rangle = {}&{}^{3i}\left\langle\lambda_k - \frac{2}{3}\mu_k\right\rangle_{LF}(\nabla\cdot{}^{3i}\langle\underline{U}_k\rangle_{LF})\underline{\underline{I}} \\
&+ {}^{3i}\langle\mu_k\rangle_{LF}[\nabla, {}^{3i}\langle\underline{U}_k\rangle_{LF} + (\nabla, {}^{3i}\langle\underline{U}_k\rangle_{LF})_c] \\
&+ {}^{3i}\left\langle\lambda_k - \frac{2}{3}\mu_k\right\rangle_{LF}{}^{3i}\langle\nabla\cdot\underline{\tilde{U}}_{kLF}\rangle\underline{\underline{I}} \\
&+ {}^{3i}\langle\mu_k\rangle_{LF}\,{}^{3i}\langle\nabla, \underline{\tilde{U}}_{kLF} + (\nabla, \underline{\tilde{U}}_{kLF})_c\rangle \\
&+ {}^{3i}\left\langle\left(\tilde{\lambda}_{kLF} - \frac{2}{3}\tilde{\mu}_{kLF}\right)\nabla\cdot\underline{\tilde{U}}_{kLF}\right\rangle\underline{\underline{I}}
\end{aligned}
$$

$$+ {}^{3i}\langle \tilde{\mu}_{kLF}[\nabla, \underline{\tilde{U}}_{kLF} + (\nabla, \underline{\tilde{U}}_{kLF})_c]\rangle$$
$$+ {}^{t3i}\left\langle\left(\lambda'_k - \frac{2}{3}\mu'_k\right)\nabla \cdot \underline{U}'_k\right\rangle \underline{\underline{I}}$$
$$+ {}^{t3i}\langle \mu'_k[\nabla, \underline{U}'_k + (\nabla, \underline{U}'_k)_c]\rangle. \qquad \text{(C.4)}$$

When λ_k and μ_k are independent of velocity gradients, ${}^{3i}\langle\lambda_k\rangle_{LF} = \lambda_k$, ${}^{3i}\langle\mu_k\rangle_{LF} = \mu_k$, $\tilde{\lambda}_{kLF} = \lambda'_k = 0$, and $\tilde{\mu}_{kLF} = \mu'_k = 0$. In addition, ${}^{3i}\langle\nabla \cdot \underline{\tilde{U}}_{kLF}\rangle = 0$, ${}^{3i}\langle\nabla, \underline{\tilde{U}}_{kLF}\rangle = 0$, and ${}^{3i}\langle(\nabla, \underline{\tilde{U}}_{kLF})_c\rangle = 0$. Consequently, for Newtonian fluids, Eq. (C.4) simplifies to

$${}^{t}\langle{}^{3i}\langle\underline{\underline{\tau}}_k\rangle\rangle = \left(\lambda_k - \frac{2}{3}\mu_k\right)(\nabla.{}^{3i}\langle\underline{U}_k\rangle_{LF})\underline{\underline{I}}$$
$$+ \mu_k[\nabla, {}^{3i}\langle\underline{U}_k\rangle_{LF} + (\nabla, {}^{3i}\langle\underline{U}_k\rangle_{LF})_c], \qquad \text{(C.5)}$$

which is precisely the result given in Eq. (5.5.4c).

APPENDIX D

Evaluation of ${}^{t}\langle {}^{2i}\langle \underline{J}_{qk}\rangle\rangle$ for isotropic conduction with variable conductivity and with $\gamma_v = 1$

The Fourier law of isotropic conduction for fluid phase k is

$$\underline{J}_{qk} = -\kappa_k \nabla T_k = -\frac{\kappa_k}{c_{vk}} c_{vk} \nabla T_k, \tag{D.1}$$

which is valid for variable conductivity. Because $\nabla u_k = c_{vk}\nabla T_k$, Eq. (D.1) can be written in a form relating the heat flux vector and the gradient of internal energy. Thus,

$$\underline{J}_{qk} = -\beta_k \nabla u_k, \tag{D.2}$$

where $\beta_k = \frac{\kappa_k}{c_{vk}}$. When κ_k or c_{vk}, or both, vary with temperature, we write [36]

$$\beta_k = {}^{2i}\langle \beta_k\rangle_{LF} + \tilde{\beta}_{kLF} + \beta'_k. \tag{D.3}$$

Accordingly,

$$\begin{aligned} {}^{2i}\langle \underline{J}_{qk}\rangle = &-{}^{2i}\langle \beta_k\rangle_{LF}(\nabla\, {}^{2i}\langle u_k\rangle_{LF} + {}^{2i}\langle \nabla \tilde{u}_{kLF}\rangle + {}^{2i}\langle \nabla u'_k\rangle) \\ &-{}^{2i}\Big\langle \tilde{\beta}_{kLF}\left(\nabla^{2i}\langle u_k\rangle_{LF} + \nabla \tilde{u}_{kLF} + \nabla u'_k\right)\Big\rangle \\ &-{}^{2i}\langle \beta'_k\left(\nabla^{2i}\langle u_k\rangle_{LF} + \nabla \tilde{u}_{kLF} + \nabla u'_k\right)\rangle . \end{aligned} \tag{D.4}$$

In deriving Eq. (D.4), the relation ${}^{2i}\langle \nabla^{2i}\langle u_k\rangle_{LF}\rangle = \nabla^{2i}\langle u_k\rangle_{LF}$ has been used. Subsequently, time averaging leads to

$$
\begin{aligned}
{}^{t}\langle {}^{2i}\langle \underline{J}_{qk}\rangle\rangle &= -{}^{2i}\langle \beta_k\rangle_{LF}(\nabla^{2i}\langle u_k\rangle_{LF} + {}^{2i}\langle \nabla\tilde{u}_{kLF}\rangle) \\
&\quad -{}^{2i}\langle \tilde{\beta}_{kLF}\nabla\tilde{u}_{kLF}\rangle - {}^{t2i}\langle \beta_k'\nabla u_k'\rangle. \qquad \text{(D.5)}
\end{aligned}
$$

When β_k is a constant, ${}^{2i}\langle \beta_k\rangle_{LF} = \beta_k$, and $\tilde{\beta}_{kLF} = \beta_k' = 0$. In addition, ${}^{2i}\langle \nabla\tilde{u}_{kLF}\rangle = 0$. Consequently, Eq. (D.5) simplifies to

$$
{}^{t}\langle {}^{2i}\langle \underline{J}_{qk}\rangle\rangle = -\left(\frac{\kappa_k}{c_{vk}}\right)\nabla^{2i}\langle u_k\rangle_{LF} = -\beta_k\nabla^{2i}\langle u_k\rangle_{LF}, \qquad \text{(D.6)}
$$

which is precisely the result given by Eq. (D.2).

APPENDIX E

Further justifications for assuming α_k', A_k', and v_k' are negligible

An additional appendix is available on the book's Web site (www.cambridge.org/9781107012950).

References

[1] W. T. Sha and J. C. Slattery, "Local Volume-Time Averaged Equations of Motion for Dispersed, Turbulent, Multiphase Flows," NUREG/CR-1491, ANL-80–51 (1980).

[2] W. T. Sha, "An Overview on Rod-Bundle Thermal-Hydraulic Analysis. *Nucl. Engi. Design*, **62**:1–24 (1980).

[3] W. T. Sha, B. T. Chao, and S. L. Soo, "Porous Media Formulation for Multiphase Flow with Heat Transfer," *Nucl. Eng. Design*, **82**:93–106 (1984).

[4] W. T. Sha, B. T. Chao, and S. L. Soo, "Averaging Procedures of Multiphase Conservation Equations," First Proceedings of Nuclear Thermal Hydraulics, 1983 Winter Meeting, October 31 to November 3, 1983, San Francisco, California, Sponsored by the Thermal Hydraulic Division of the American Nuclear Society.

[5] W. T. Sha, B. T. Chao, and S. L. Soo, "Time Averaging of Volume Averaged Conservation Equations of Multiphase Flow," *AIChE Symposium Series*, No. 255, pp. 420–426, Heat Transfer Seattle (1983).

[6] W. T. Sha and B. T. Chao, "Novel Porous Media Formulation for Multiphase Flow Conservation Equations," *Nucl. Eng. Design*, **237**:918–942 (2007).

[7] S. Whitaker, "Advances in Theory of Fluid Motion in Porous Media," *Ind. Eng. Chem.*, **61**:14–28 (1969).

[8] J. C. Slattery, "Flow of Viscoelastic Fluids through Porous Media," *AIChE J.*, **13**:1066–1071 (1967).

[9] S. L. Soo, "Dynamics of Multiphase Flow," *I&EC Fundam.* **4**:425–433 (1965).

[10] S. Whitaker, "Diffusion and Dispersion in Porous Media," *AIChE J.*, **13**:420–427 (1967).

[11] T. B. Anderson and R. Jackson, "A Fluid Mechanical Description of Fluidized Beds," *I&EC Fundam.*, **6**:527–539 (1967).

[12] J. C. Slattery, "Multiphase Viscoelastic Flow through Porous Media," *AIChE J.*, **14**:50 (1968).

[13] G. B. Wallis, *One-Dimensional Two-Phase Flow*, McGraw-Hill, New York (1969).

[14] J. C. Slattery, "Two-Phase Flow through Porous Media," *AIChE J.*, **16**:345 (1970).

[15] D. A. Drew, "Averaged Field Equations for Two-Phase Media," *Studies Appl. Math.*, **50**:133–166 (1971).

[16] J. G. Patel, M. G. Hegde, and J. C. Slattery, "Further Discussion of Two-Phase Flow in Porous Media," *AIChE J.*, **18**:1062 (1972).

[17] Y. Bachmat, "Spatial Macroscopization of Processes in Heterogeneous Systems," *ISR. J. Technol.*, **10**:391 (1972).

[18] J. Bear, *Dynamics of Fluids in Porous Media*, American Elsevier, New York (1972).

[19] S. Whitaker, "The Transport Equations for Multiphase Systems," *Chem. Eng. Sci.*, **28**:139 (1973).

[20] S. W. Hopke and J. C. Slattery, "Bounding Principles for Two-Phase Flow Systems," *Int. J. Multiphase Flow*, **1**:727 (1975).

[21] W. G. Gray, "A Derivation of the Equations for Multiphase Transport," *Chem. Eng. Sci.*, **30**:229 (1975).

[22] W. G. Gray and K. O'Neill, "On the General Equations for Flow in Porous Media and Their Reduction to Darcy's Law," *Water Resour. Res.*, **12**:148 (1976).

[23] W. G. Gray and P. C. Y. Lee, "On the Theorems for Local Volume Averaging of Multiphase Systems," *Int. J. Multiphase Flow*, **3**:333–340 (1977).

[24] J. M. Delhaye and J. L. Archard, "On the Averaging Operators Introduced in Two-Phase Flow Modeling," OECD/NEA, Specialists' Meeting on Transient Two-Phase Flow, Toronto, Canada, August 3–4, 1976, as quoted in [26].

[25] J. M. Delhaye, "Instantaneous Space-Averaged Equations," in *Two-Phase Flows and Heat Transfer*, Vol. 1, edited by S. Kakac and F. Mayinger, p. 81, Hemisphere, Washington, DC (1977).

[26] J. M. Delhaye, "Local Time-Averaged Equations," in *Two-Phase Flows and Heat Transfer*, Vol. 1, edited by S. Kakac and F. Mayinger, p. 101, Hemisphere, Washington, DC (1977).

[27] F. K. Lehner, "A Derivation of the Field Equations for Slow Viscous Flow through a Porous Media," *Ind. Eng. Chem. Fundam.*, **18**(1), 41–45 (1979).

[28] A. J. Reynolds, "Reviews–Thermo-Fluid Dynamic Theory of Two-Phase Flow, by M. Ishii," *J. Fluid Mech.*, **78**(3):638–640 (1976).

[29] M. Ishii, *Thermo-Fluid Dynamic Theory of Two-Phase Flow*, Eyrolles, Paris (1975).

[30] M. Ishii and T. Hibiki, *Thermo-Fluid Dynamics of Two-Phase Flow*, Springer Science, New York (2006).

[31] W. T. Sha and S. L. Soo, "On the Effect of Term in Multiphase Mechanics," *Int. J. Multiphase Flow*, **5**:153–158 (1979).

[32] S. L. Soo, "Equations of Multiphase-Multidomain Mechanics," in *Multiphase Transport*, Vol. 1, edited by

T. Veziroglu, pp. 291–305, Hemisphere, Washington, DC (1980).

[33] I. J. Campbell and A. S. Pitcher. "Shock Waves in Liquid Containing Gas Bubbles," *Proc. Roy. Soc. Ser. A*, **243**:534 (1958).

[34] J. P. Hinz, *Turbulence*, McGraw-Hill NY (1978).

[35] S. Winnikow and B. T. Chao, "Droplet Motion in Purified Systems," *Phys. Fluids*, **9**:50–61 (1966).

[36] W. T. Sha, B. T. Chao, and S. L. Soo, "Time and Volume-Averaged Conservation Equations for Multiphase Flow, Part 1: System without Internal Solid Structures," NUREG/CR-3989, ANL-84–66 (December 1984).

[37] W. T. Sha, H. M. Domanus, R. C. Schmitt, J. J. Oras, and E. J. H. Lin, "COMMIX-1: A Computer Program for Three-Dimension, Transient, Single-Phase Thermal-Hydraulic Analysis," NUREG/CR-0785, ANL-77–96 (September 1978); H. M. Domanus, R. C. Schmitt, W. T. Sha, and V. L. Shah, "COMMIX-1A: A Three-Dimensional, Transient, Single-Phase Computer Program for Thermal-Hydraulic Analysis of Single and Multi-Component Systems," NUREG/CR-2896, ANL-82–25 (December 1983); W. T. Sha, F. F. Chen, H. M. Domanus, C. C. Miao, R. C. Schmitt, and V. L. Shah, "COMMIX-1B: A Three-Dimensional, Transient, Single-Phase Computer Program for Thermal Hydraulic Analysis of Single and Multicomponent Systems," NUREG/CR-4348, ANL-85–42 (September 1985); H. M. Domanus, Y. S. Cha, T. H. Chien, R. C. Schmitt, and W. T. Sha, "COMMIX-1C: A Three-Dimensional, Steady-State/Transient, Single-Phase Computer Program for Thermal-Hydraulic Analysis of Single/Multicomponent Systems," NUREG/CR-5649, ANL-90–33 (September 1990); and T. H. Chien, J. G. Sun, and W. T. Sha, "A Preliminary Compilation of the New

Models and the Input Description for COMMIX-1D: A Three Dimensional, Transient, Single Phase Multicomponent Computer Code for Containment Thermal Hydraulic Analysis," Transmittal Letter from W. T. Sha (ANL) to Allen Notafrancesco (U.S.N.R.C.-Research) (July 14, 1995).

[38] R. W. Davis, E. F. Moore, and L. P. Purtell, "A Numerical-Experiment Study of Confined Flow around Rectangular Cylinders," *Phys. Fluids*, **27**:46–59 (1984).

[39] U. Ghia, K. N. Ghia, and C. T. Shin, "High-*Re* Solutions for Incompressible Flow Using the Navier-Stokes Equations and a Multigrid Method," *J. Compu. Phys.*, **48**:387–411 (1982).

[40] K. Sakai, J. G. Sun, and W. T. Sha, "Implementation of the High-Order Schemes Quick and LECUSSO in the COMMIX-IC Program," ANL/ATHRP-47 (February 1994).

[41] B. P. Leonard, "The QUICK Finite Difference Method for the Convective-Diffusion Equation in Advances," in *Computer Methods for Partial Differential Equations, III*, IMACS Lehgh University Bethlehem, PA, USA June 20–22, 1979 (1979).

[42] M. Chapman, "FRAM-Nonlinear Damping Algorithm for the Continuity Equation," *J. Comp. Phys.*, **44**:84–103 (1981).

[43] R. McGriff, J. Chao, B. Chexal, and W. Layman, "Analysis of Mixing in Cold Leg and Downcomer of a W-Three Loop Plant, during a SBLOCA," NSAC/62, EPRI (May 1984).

[44] B. C-J. Chen, B. K. Cha, and W. T. Sha, "COMMIX-1A Analysis of Fluid and Thermal Mixing in a Model Cold Leg and Downcomer of a PWR," EPRI, NP-3557 (June 1984).

[45] W. T. Sha, C. I. Yang, T. T. Kao, and S. M. Cho, "Multi-Dimensional Numerical Modeling of Heat Exchangers," *J. Heat Transfer*, **104**:417–425 (1982).

[46] W. T. Sha, "Numerical Modeling of Heat Exchangers," *Handbook for Heat and Mass Transfer, Vol. 1, Heat Transfer Operations*, pp. 815–852, Gulf, Houston, Texas (1986).

[47] W. L. Baumann, H. M. Domanus, and W. T. Sha, "EBR-II In-Vessel Thermal-Hydraulic Transient Simulation Using the COMMIX-1A Computer Code," *Trans. Am. Nucl. Soc.*, **43**:499–501 (November 1982).

[48] W. T. Sha, "In-Vessel Thermal Hydraulic Analysis," in Proceedings of the IAHR Specialists Meeting on Liquid Metal Thermal Hydraulics in Plena and Pipes, Sunnyvale, California (January 1983).

[49] W. T. Sha and V. L. Shah, "Natural Convention Phenomena in a Prototypics PWR during a Postulated Degraded Core Accident," EPRI TR-103574 (January 1994).

[50] W. A. Stewart, A. T. Pirvzynski, and V. Svinivas, "Experiments on Natural Circulation Flows in Steam Generators during Severe Accidents," Westinghouse Report 85-5JO-RCIRC-P3 (December 1985). Also in Proceedings of International ANS-ENS Topical Meeting on Thermal Reactor Safety, San Diego, California (February 2–6, 1986), pp. XIX.6-1–XIX.6-8.

[51] W. T. Sha, T. H. Chien, J. G. Sun, and B. T. Chao, "Analysis of Large-Scale Tests for AP-600 Passive Containment Cooling System," *Nucl. Eng. and Design*, **232**:197–216 (2004).

[52] D. R. Spencer, J. Woodcook, R. F. Wright, J. E. Schmidt, M. Panes, and D. E. Christenson, "Reactor Passive Containment Cooling Systems Tests Scaling Evaluation and

Analysis," in Proceedings of ASME/JSME Nuclear Engineering Conference, San Francisco, California (March 21–24, 1993).

[53] J. J. Ha, and T. Aldemir, "Thermal-Hydraulic Analysis of the OSURR Pool for Power Upgrade with Natural Convection Cooling," in Proceedings of the 1986 International RERTR Meeting, ANL/RERTR/TM-9, 473–489, Argonne National Laboratory (1988).

[54] W. T. Sha, "Generic Casting Modeling," in Proceedings of the International Meeting on Processes Science, Conf-87102271-1, NATO Advanced Study Institute, Cesame, Turkey (October 6–7, 1987).

[55] H. M. Domanus, R. C. Schmitt, W. T. Sha, E. M. Petrill, W. L. Howe, and J. P. Bass, "Computer Simulation of Mixing Phenomena in AFBC Boiler," in Proceedings of the Atmospheric Fluidized Combustion Seminar, CONF-8604 137-1, Palo Alto, California (April 7–10, 1986).

[56] J. R. Hull, K. V. Liu, W. T. Sha, J. Komal, and C. E. Nielson, "Dependence of Ground Heat Loss upon the Solar Pond Size and Parameter Insulation: Calculation and Experimental Results," *J. Solar Energy*, **33**:25–33 (1984).

[57] W. T. Sha and S. L. Soo, "Modeling and Optimization of Flotation Processes for Coal Precleaning," ANL/FE-90/4 (September 1990).

[58] C. E. Brennen, *Fundamentals of Multiphase Flow*, Cambridge University Press New York, NY (2005).

[59] G. H. Crapiste, E. Rotstein, and S. Whitaker, "A General Closure Scheme for Method of Volume Averaging," *Chem. Eng. Sci.*, **41**:227–235 (1986).

[60] A. K. M. Fazle Hussain, "Role of Coherent Structure in Turbulent Shear Flow," *Proc. Indian Acad. Sci.*, **4**(2):129–175 (1981).

[61] C. Crowe, M. Sommerfeld, and Y. Tsuji, *Multiphase Flows with Droplets and Particles*, CRC Press, Boca Raton, Florida (1998).

[62] N. I. Kolev, *Multiphase Flow Dynamics*, 3rd edition Springer Berlin, NY: Springer (2007).

[63] A. Padilla Jr., and D. S. Rowe, "A Donor Flow Formulation for Momentum Flux Differencing," *Trans. Am. Nucl. Sci.*, **46**:851–852 (1984).

[64] W. T. Sha and Joe F. Thompson, "Rod-Bundle Thermal-Hydraulic Analysis Using Boundary-Fitted Coordinates System," NUREG/CR-0001, ANL-78-1 (January, 1979).

[65] B. C.-J. Chen, T. H. Chien, and W. T. Sha, BODYFIT-2PE-HEM: LWR Core Thermal-Hydraulic Code Using Boundary-Fitted Coordinates and Two-Phase Homogeneous Equilibrium Model, Vol. 1, Theory and Formulations, Vol 2, User's Manual, and Vol. 3, Validations and Applications. EPRI NP-3768-CCM, Project 1383-1 Computer Code Manual (August, 1985).

[66] W. T. Sha, "Fluid Flow and Heat Transfer in Arbitrary Three-Dimensional Geometries via Boundary-Fitted Coordinates Transformation," Published in Special Edition of Fifteenth Anniversary of Engineering Thermophysics Institute, Xian Jiangtong University, Xian, China (September 19–21, 1994).

[67] L. W. Florschaetz and B. T. Chao, "On the Mechanics of Vapor Bubble Collapse," *Journal of Heat Transfer, Trans. ASME*, 87C:209–220 (1965).

Index

Printed in the United States
By Bookmasters